Organic Synthesis:
The Disconnection Approach

Organic Synthesis:
The Disconnection Approach
2nd Edition

Stuart Warren

Chemistry Department, Cambridge University, UK

and

Paul Wyatt

School of Chemistry, University of Bristol, UK

A John Wiley and Sons, Ltd., Publication

This edition first published 2008
© 2008 John Wiley & Sons, Inc.

Registered office
John Wiley & Sons Ltd, The Atrium, Southern Gate, Chichester, West Sussex, PO19 8SQ, United Kingdom

For details of our global editorial offices, for customer services and for information about how to apply for permission to reuse the copyright material in this book please see our website at www.wiley.com.

Library of Congress Cataloging-in-Publication Data

Warren, Stuart G.
  Organic synthesis : the disconnection approach / Stuart Warren and Paul Wyatt.—2nd ed.
    p. cm.
  Includes bibliographical references and index.
  ISBN 978-0-470-71237-5 (cloth)—ISBN 978-0-470-71236-8 (pbk. :
  alk. paper)
  1. Organic compounds–Synthesis. I. Wyatt, Paul. II. Title.
  QD262.W284 2008
  547′.2–dc22

                                        2008033269

A catalogue record for this book is available from the British Library.

ISBN 978-0-470-7-12375 (HBK)   978-0-470-7-12368 (PBK)

Typeset in 10/12 Times-Roman by Laserwords Private Limited, Chennai, India

The first edition was written with the active participation of Denis Marrian who died in 2007. We dedicate this second edition to Denis Haigh Marrian, 1920–2007, a great teacher and friend.

Contents

Preface

In the 26 years since Wiley published *Organic Synthesis: The Disconnection Approach* by Stuart Warren, this approach to the learning of synthesis has become widespread while the book itself is now dated in content and appearance. In 2007, Wiley published *Organic Synthesis: Strategy and Control* by Paul Wyatt and Stuart Warren. This much bigger book is designed as a sequel for fourth year undergraduates and research workers in universities and industry. The accompanying workbook was published in 2008. This new book made the old one look very dated in style and content and exposed gaps between what students were expected to understand in the 1980s and what they are expected to understand now. This second edition is intended to fill some of those gaps.

The plan of the original book is the same in the second edition. It alternates chapters presenting new concepts with strategy chapters that put the new work in the context of overall planning. The 40 chapters have the same titles: some chapters have hardly been changed while others have undergone a thorough revision with considerable amounts of new material. In most cases examples from recent years are included.

One source of new material is the courses that the authors give in the pharmaceutical industry. Our basic course is 'The Disconnection Approach' and the material we have gathered for this course has reinforced our attempts to give reasons for the synthesis of the various compounds which we believe enlivens the book and makes it more interesting for students. We hope to complete a second edition of the workbook shortly after the publication of the main text.

The first edition of the textbook was in fact the third in a series of books on organic chemistry published by Wiley. The first: *The Carbonyl Group: an Introduction to Organic Mechanisms*, published in 1974, is a programmed book asking for a degree of interaction with the reader who was expected to solve problems while reading. People rarely use programmed learning now as the method has been superseded by interactive programmes on computers. Paul Wyatt is writing an electronic book to replace *The Carbonyl Group* which will complete a package of an electronic book and books with associated workbooks in a uniform format that we hope will prove of progressive value as students of organic chemistry develop their careers.

Stuart Warren and Paul Wyatt
March 2008.

General References

Full details of important books referred to by abbreviated titles in the chapters to avoid repetition.

Clayden, *Lithium*: J. Clayden, *Organolithiums: Selectivity for Synthesis*, Pergamon, 2002.

Clayden, *Organic Chemistry*: J. Clayden, N. Greeves, S. Warren and P. Wothers, *Organic Chemistry*, Oxford University Press, Oxford, 2000.

Comp. Org. Synth.: eds. Ian Fleming and B. M. Trost, *Comprehensive Organic Synthesis*, Pergamon, Oxford, 1991, six volumes.

Corey, *Logic*: E. J. Corey and X.-M. Cheng, *The Logic of Chemical Synthesis*, Wiley, New York, 1989.

Fieser, *Reagents:* L. Fieser and M. Fieser, *Reagents for Organic Synthesis*, Wiley, New York, 20 volumes, 1967–2000, later volumes by T.-L. Ho.

Fleming, *Orbitals*: Ian Fleming, *Frontier Orbitals and Organic Chemical Reactions*, Wiley, London, 1976.

Fleming, *Syntheses*: Ian Fleming, *Selected Organic Syntheses*, Wiley, London, 1973.

Houben-Weyl; *Methoden der Organischen Chemie*, ed. E. Müller, and *Methods of Organic Chemistry*, ed. H.-G. Padeken, Thieme, Stuttgart, many volumes 1909–2004.

House: H. O. House, *Modern Synthetic Reactions*, Benjamin, Menlo Park, Second Edition, 1972.

Nicolaou and Sorensen: K. C. Nicolaou and E. Sorensen, *Classics in Total Synthesis: Targets, Strategies, Methods*. VCH, Weinheim, 1996. Second volume now published.

Saunders, *Top Drugs*: J. Saunders, *Top Drugs: Top Synthetic Routes*, Oxford University Press, Oxford, 2000.

Strategy and Control; P. Wyatt and S. Warren, *Organic Synthesis: Strategy and Control*, Wiley, Chichester, 2007 and *Workbook*, 2008.

Vogel: B. S. Furniss, A. J. Hannaford, P. W. G. Smith, and A. R. Tatchell, *Vogel's Textbook of Practical Organic Chemistry*, Fifth Edition, Longman, Harlow, 1989.

1 The Disconnection Approach

This book is about making molecules. Or rather it is to help you design your own syntheses by logical and sensible thinking. This is not a matter of guesswork but requires a way of thinking backwards that we call the disconnection approach.

When you plan the synthesis of a molecule, all you know for certain is the structure of the molecule you are trying to make. It is made of atoms but we don't make molecules from atoms: we make them from smaller molecules. But how to choose which ones? If you wanted to make, say, a wooden joint, you would look in a do-it-yourself book on furniture and you would find an 'exploded diagram' showing which pieces you would need and how they would fit together.

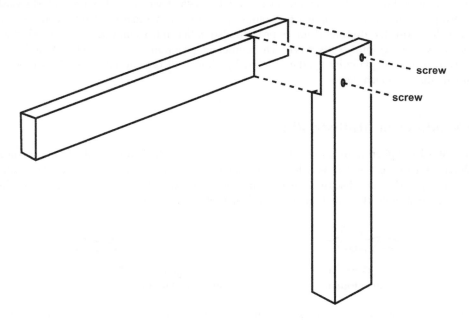

The disconnection approach to the design of synthesis is essentially the same: we 'explode' the molecule into smaller starting materials on paper and then combine these by chemical reactions. It isn't as easy as making wooden joints because we have to use logic based on our chemical knowledge to choose these starting materials. The first chemist to suggest the idea was Robert Robinson who published his famous tropinone synthesis[1] in 1917. His term was 'imaginary

Organic Synthesis: The Disconnection Approach. Second Edition   Stuart Warren and Paul Wyatt
© 2008 John Wiley & Sons, Ltd

hydrolysis' and he put dashed lines across a tropinone structure.

Tropinone: Robinson's Analysis

This was a famous synthesis because it is so short and simple and also because it makes a natural product in a way that imitates nature. The reaction is carried out at pH 7 in water. In fact Robinson didn't use acetone, as suggested by his 'imaginary hydrolysis', but acetone dicarboxylic acid. This procedure is an improved one invented by Schöpf[2] in 1935.

Tropinone: Synthesis

Amazingly, nobody picked up the idea until the 1960s when E. J. Corey at Harvard was considering how to write a computer program to plan organic syntheses.[3] He needed a systematic logic and he chose the disconnection approach, also called retrosynthetic analysis. All that is in this book owes its origin to his work. The computer program is called LHASA and the logic survives as a way of planning syntheses used by almost all organic chemists. It is more useful to humans than to machines.

The Synthesis of Multistriatin

Multistriatin **1** is a pheromone of the elm bark beetle. This beetle distributes the fungus responsible for Dutch elm disease and it was hoped that synthetic multistriatin might trap the beetle and prevent the spread of the disease. It is a cyclic compound with two oxygen atoms both joined to the same carbon atom (C-6 in **1**) and we call such ethers *acetals*.

1; multistriatin   acetal functional group

We know one good way to make acetals: the reliable acid-catalysed reaction between two alcohols or one diol and an aldehyde or ketone.

Intending to use this reliable reaction for our acetal we must disconnect the two C–O bonds to C-6 and reveal the starting material **2**, drawn first in a similar way to **1**, and then straightened out to look more natural **2a**. Numbering the carbon atoms helps to make sure **2** and **2a** are the same.

We now have a continuous piece of carbon skeleton with two OH groups and a ketone. No doubt we shall make this by forming a C–C bond. But which one? We know that ketones can form nucleophilic enolates so disconnecting the bond between C–4 and C–5 is a good choice because one starting material **3** is symmetrical. As we plan to use an enolate we need to make **3** nucleophilic and therefore **4** must be electrophilic so we write plus and minus charges to show that.

Anion **3** can be made from the available ketone **5** but the only sensible way to make **4** electrophilic is to add a leaving group X, such as a halogen, deciding later exactly what to use.

Compound **6** has three functional groups. One is undefined but the other two must be alcohols and must be on adjacent carbon atoms. There is an excellent reaction to make such a combination: the dihydroxylation of an alkene with a hydroxylating agent such as OsO$_4$. A good starting material becomes the unsaturated alcohol **7a** as that is known.

In one synthesis[4] the alcohol **7a** was made from the available acid **8** and the leaving group (X in **6**) was chosen as tosylate (OTs; toluene-*p*-sulfonate).

The two pieces were joined together by making the enolate of **5** and reacting it with **7**; X = OTs. The unsaturated ketone **9** was then oxidised with a peroxyacid to give the epoxide **10** and cyclisation with the Lewis acid SnCl$_4$ gave the target molecule (TM) multistriatin **1**.

You may have noticed that the synthesis does not exactly follow the analysis. We had planned to use the keto-diol **2b** but in the event this was a less practical intermediate than the keto-epoxide **10**. It often turns out that experience in the laboratory reveals alternatives that are better than the original plan. The basic idea—the strategy—remains the same.

Summary: Routine for Designing a Synthesis

1. *Analysis*
   (a) Recognise the functional groups in the target molecule.
   (b) Disconnect with known reliable reactions in mind.
   (c) Repeat as necessary to find available starting materials.
2. *Synthesis*
   (a) Write out the plan adding reagents and conditions.
   (b) Modify the plan according to unexpected failures or successes in the laboratory.

We shall develop and continue to use this routine throughout the book.

What the Rest of the Book Contains

The synthesis of multistriatin just described has one great fault: no attempt was made to control the stereochemistry at the four chiral centres (black blobs in **11**). Only the natural stereoisomer attracts the beetle and stereoselective syntheses of multistriatin have now been developed.

**11; chiral centres
in multistriatin**

We must add stereochemistry to the list of essential background knowledge an organic chemist must have to design syntheses effectively. That list is now:

1. An understanding of reaction mechanisms.
2. A working knowledge of reliable reactions.
3. An appreciation that some compounds are readily available.
4. An understanding of stereochemistry.

Don't be concerned if you feel you are weak in any of these areas. The book will strengthen your understanding as you progress. Each chapter will build on whichever of the four points are relevant. If a chapter demands the understanding of some basic chemistry, there is a list of references at the start to chapters in Clayden *Organic Chemistry* to help you revise. Any other textbook of organic chemistry will have similar chapters.

The elm bark beetle pheromone contains three compounds: multistriatin, the alcohol **12** and α-cubebene **13**. At first we shall consider simple molecules like **12** but by the end of the book we shall have thought about molecules at least as complex as multistriatin and cubebene.

1; multistriatin   12   13; α-cubebene

Multistriatin has been made many times by many different strategies. Synthesis is a creative science and there is no 'correct' synthesis for a molecule. We shall usually give only one synthesis for each target in this book: you may well be able to design shorter, more stereochemically controlled, higher yielding, more versatile—in short better—syntheses than those already published. If so, you are using the book to advantage.

References

1. R. Robinson, *J. Chem. Soc.*, 1917, **111**, 762.
2. C. Schöpf and G. Lehmann, *Liebig's Annalen*, 1935, **518**, 1.
3. E. J. Corey, *Quart. Rev.*, 1971, **25**, 455; E. J. Corey and X.-M. Cheng, *The Logic of Chemical Synthesis*, Wiley, New York, 1989.
4. G. T. Pearce, W. E. Gore and R. M. Silverstein, *J. Org. Chem.*, 1976, **41**, 2797.

2 Basic Principles: Synthons and Reagents Synthesis of Aromatic Compounds

Background Needed for this Chapter References to Clayden, *Organic Chemistry:*
Electrophilic aromatic substitution; chapter 22. (Electrophilic Aromatic Substitution)
Nucleophilic aromatic substitution: chapter 23 (Electrophilic Alkenes).
Reduction: chapter 24 (Chemoselectivity: Selective Reactions and Protection).

Synthesis of Aromatic Compounds

The benzene ring is a very stable structural unit. Making aromatic compounds usually means adding something(s) to a benzene ring. The disconnection is therefore almost always of a bond joining a side chain to the benzene ring. All we have to decide is when to make the disconnection and which reagents to use. You will meet the terms *synthon* and *functional group interconversion* (FGI) in this chapter.

Disconnection and FGI

You already know that disconnections are the reverse of known reliable reactions so you should not make a disconnection unless you have such a reaction in mind. In designing a synthesis for the local anaesthetic benzocaine **1**, we see an ester group and know that esters are reliably made from some derivative of an acid (here **2**) and an alcohol (here ethanol). We should disconnect the C–O ester bond. From now on we will usually write the reason for a disconnection or the name of the forward reaction above the arrow.

The sign for a disconnection on a molecule is some sort of wiggly line across the bond being disconnected. You can draw this line in any way you like within reason. The 'reaction arrow' is the 'implies' arrow from logic. The argument is that the existence of any ester *implies* that it can be made from an acid and an alcohol.

Organic Synthesis: The Disconnection Approach. Second Edition Stuart Warren and Paul Wyatt
© 2008 John Wiley & Sons, Ltd

We should now like to disconnect either the NH_2 or the CO_2H group but we know of no good reactions corresponding to those disconnections. We need to change both groups into some other groups that can be added to a benzene ring by a known reliable reaction. This process is called *functional group interconversion* or FGI for short and is an imaginary process, just like a disconnection. It is the reverse of a real reaction. Here we know that we can make amino groups by reduction of nitro groups and aryl carboxylic acids by oxidation of alkyl groups. The FGIs are the reverse of these reactions.

We 'oxidised' the amino group first and 'reduced' the acid second. The order is unimportant but is something we come back to in the forward reaction. What matters is that we have found a starting material **4** that we know how to make. If we disconnect the nitro group **4a** we shall be left with toluene **5** and toluene can be nitrated in the *para*-position with a mixture of nitric and sulfuric acids.

Now we should write out the synthesis. You cannot of course predict exactly which reagents and conditions will be successful and no sensible organic chemist would attempt to do this without studying related published work. It is enough to make suggestions for the type of reagent needed. We shall usually give the reagents used in the published work and conditions where they seem to matter. Here it is important to nitrate first and oxidise second to get the right substitution pattern.[1]

Synthons Illustrated by Friedel-Crafts Acylation

The useful disconnection **6a** corresponds to Friedel-Crafts acylation of aromatic rings and is the obvious one on the ketone **6** having the perfume of hawthorn blossom. Reaction[2] of ether **7** with MeCOCl and $AlCl_3$ gives **6** in 94–96% yield—a good reaction indeed.

In both this reaction and the nitration of toluene we used to make benzocaine, the reagent is a cation: $MeCO^+$ for the Friedel-Crafts and NO_2^+ for the nitration. Our first choice on disconnecting a bond to a benzene ring is to look for a cationic reagent so that we can use electrophilic aromatic substitution. We know not only which bond to break but also in which sense electronically to break it. In principle we could have chosen either polarity from the same disconnection: **a** (we actually chose) or **b** (we did not).

The four fragments **8–11** are *synthons* — that is idealised ions that may or may not be involved in the actual reaction but help us to work out which reagent to choose. As it happens, synthon **11** is a real intermediate but the others are not. For an anionic synthon like **10** the reagent is often the corresponding hydrocarbon as H^+ is lost during the reaction. For a cationic synthon like **11** the reagent is often the corresponding halide as that will be lost as a leaving group during the reaction. It is a matter of personal choice in analysing a synthesis problem whether you draw the synthons or go direct to the reagents. As you become more proficient at retrosynthetic analysis, you will probably find that drawing the synthons becomes unnecessary and cumbersome.

Synthons Illustrated by Friedel-Crafts Alkylation

Friedel-Crafts alkylation is also useful though less reliable than acylation. With that in mind, we could disconnect BHT **13** ('Butylated Hydroxy-Toluene') at either bond **b** to remove the methyl group or bond **a** to remove both *t*-butyl groups. There are various reasons for preferring **a**. *para*-Cresol **15** is available whereas **14** is not. The *t*-butyl cation is a much more stable intermediate than the methyl cation — and *t*-alkylations are among the most reliable. Finally the OH group is more powerfully *ortho*-directing than the methyl group.

We have a choice of reagents for the *t*-butyl cation: a halide with Lewis acid catalysis, and *t*-butanol or isobutene with protic acid catalysis. The least wasteful is the alkene as nothing is lost. Protonation gives the *t*-butyl cation and two *t*-butyl groups are added in one operation.[3]

Functional Group Addition Illustrated by Friedel-Crafts Alkylation

Attempting Friedel-Crafts alkylation with primary halides often gives the 'wrong' product by rearrangement of the intermediate cation. If we want to make *i*-butylbenzene **16**, it seems obvious that we should alkylate benzene with an *i*-butyl halide, e.g. **18** and AlCl$_3$.

This reaction gives two products **21** and **22** but neither contains the *i*-butyl group. Both contain instead the *t*-butyl group. The intermediate complex rearranges by hydride shift **19** into the *t*-butyl cation **20** as the primary cation **17** is too unstable.

Polyalkylation was an advantage in the synthesis of BHT **13**: it is the rearrangement that is chiefly unacceptable here. Friedel-Crafts acylation avoids both problems. The acyl group does not rearrange and the product is deactivated towards further electrophilic attack by the electron-withdrawing carbonyl group. We have an extra step: reduction of the ketone to a CH$_2$ group. There are various ways to do this (see chapter 24)—here the Clemmensen reduction is satisfactory.[4]

The preliminary to the corresponding disconnection is the 'addition' (imaginary) of a functional group where there was none. We call this FGA (functional group addition). The corresponding known reliable reaction is the removal of the functional group. We could put the carbonyl group anywhere but we put it next to the benzene ring as it then allows us to do a reliable disconnection.

Reliable Reagents for Electrophilic Substitution

Table 2.1 summarises the various reagents we have mentioned (and some we haven't). Full details of mechanisms, orientation and applications appear in *Clayden* chapter 22.

TABLE 2.1   Reagents for aromatic electrophilic substitution

Synthon	Reagent	Reaction	Comments
R^+	$RBr + AlCl_3$	Friedel-Crafts	good for t-alkyl
	ROH or alkene $+H^+$	alkylation[5]	OK for s-alkyl
RCO^+	$RCOCl + AlCl_3$	Friedel-Crafts acylation	very general
NO_2^+	$HNO_3 + H_2SO_4$	nitration	very vigorous
Cl^+	$Cl_2 + FeCl_3$	chlorination	other Lewis acids used too
Br^+	$Br_2 + Fe\ (=FeBr_3)$	bromination	other Lewis acids used too
$^+SO_2OH$	H_2SO_4	sulfonation	may need fuming H_2SO_4
$^+SO_2Cl$	$ClSO_2OH + H_2SO_4$	chloro-sulfonation	very vigorous
ArN_2^+	$ArNH_2 + HONO$	diazo-coupling	product is $Ar^1N=NAr^2$

Changing the Polarity: Nucleophilic Aromatic Substitution

If we make the same disconnections as before **25** and **27** but change the polarity we need electrophilic aromatic rings and nucleophilic reagents. We shall need a leaving group X (might be a halogen) on the aromatic ring **26** and reagents such as alkoxides or amines.

The nucleophilic reagents are behaving normally for alcohols or amines but the aromatic electrophiles present a problem. Benzene rings are nucleophilic, if weakly, but are not electrophilic at all. There is no S_N2 reaction on an aryl halide. To get the reactions we want, we must have *ortho-* or *para*-electron-withdrawing groups such as NO_2 or $C=O$ to accept the electrons as the nucleophile adds **28** to form **29**.

Fortunately, nitro groups go in the right positions (i.e. *ortho* and *para* but not *meta*) by direct nitration of, say, chlorobenzene. So we can be guided in our choice of polarity by the nature of the target molecule. The Lilly pre-emergent herbicide trifluralin B **31** has three electron-withdrawing groups: two nitro and one CF_3, *ortho-* and *para-* to the amine, ideal for nucleophilic substitution on **32**. The nitro groups can be introduced by nitration as Cl directs *ortho, para* while CF_3 directs *meta*.

The synthesis[5] is simplicity itself, as the synthesis of any agrochemical must be. The base in the second step is to remove the HCl produced in the reaction, not to deprotonate the amine.

Thinking Mechanistically

It is obvious that the choice between nucleophilic and electrophilic substitution must be mechanistically made but this is generally true of the choice of all disconnections, synthons and reagents. The formation of **31** was easy because the aryl chloride was activated by three groups. In the synthesis of fluoxetine (Prozac), a rather widely taken anti-depressant, aryl ether **34** is an essential intermediate.[6] Though disconnection **b** looks attractive, as a simple S_N2 reaction should work well, disconnection **a** was preferred because **34** must be a single enantiomer and enantiomerically pure alcohol **36** was available.

You should have been surprised to see fluoride as the leaving group. Fluoride is the worst leaving group among the halogens as the C–F bond is very strong: it is rare to see an S_N2 reaction with fluoride as the leaving group. Yet it is the best choice for nucleophilic aromatic substitution especially when the ring is only weakly activated as here with just one CF_3 group. In this two-step reaction, the difficult step is the addition of the nucleophile: aromaticity is destroyed and the intermediate is an unstable anion. The second step **38** is fast. Fluorine accelerates the first step as it is so electronegative and it doesn't matter that it hinders the second step as that is fast anyway.

You may have noticed something else. The formation of trifluralin **31** showed that amines are good nucleophiles for nucleophilic aromatic substitution and the nucleophile here is an amino-alcohol **36**. Direct reaction with **36** might lead to the formation of an amine instead of an ether. To avoid this, **36** is first treated with NaH to make the oxyanion and then added to **35**. The alcohol is less nucleophilic but the oxyanion is more nucleophilic than the amine. We hope you now see why an understanding of reaction mechanisms is an essential preliminary to the designing of syntheses.

Changing the Polarity: Nucleophilic Aromatic Substitution by the S_N1 Mechanism

Though the S_N2 mechanism is not available for aromatic nucleophilic substitutions, the S_N1 is providing we use the very best leaving group available. This is a molecule of nitrogen released from a diazonium salt **42** on gentle warming. A standard sequence is nitration of an aromatic compound **39** to give **40**, reduction to the amine **41** and diazotisation with $NaNO_2$/HCl to give the diazonium salt **42**. Nitrous acid HONO is the true reagent giving NO^+ that attacks at nitrogen.

The diazonium salt **42** is stable at 0–5 °C but decomposes to N_2 and an unstable aryl cation **43** on warming to room temperature. The empty orbital of **43** is in an sp^2 orbital in the plane of the aromatic ring, quite unlike the normal p orbital for cations like **20**. Reaction occurs with any available nucleophile, even water, and this is a route to phenols **45**.

This route is particularly valuable for substituents that cannot easily be added by electrophilic substitution such as OH or CN. Table 2.2 gives you a selection of reagents. For the addition of CN, Cl or Br, copper (I) derivatives usually give the best results. So the aryl nitrile **46** might come from amine **47** via a diazonium salt and routine disconnections lead us back to toluene.

The synthesis is straightforward.[7] In the laboratory you would not have to carry out the first two steps as the amine **47** can be bought. Industry makes it on a large scale by this route. Notice that we do not draw the diazonium salt. You can if you want, but it is usual to show two steps carried out without isolation of the intermediate in this style: 1. reagent A, 2. reagent B. This makes it clear that all the reagents are not just mixed together. Another style is used in Table 2.2: the reactive intermediate is in square brackets. But it is helpful to show conditions for the diazotisation as temperature control is important.

TABLE 2.2   Reagents for aromatic nucleophilic substitution on ArN_2^+

Synthon	Reagents	Comments
–OH	water	probably S_N1
–OR	alcohol ROH	probably S_N1
–CN	Cu(I)CN	may be a radical reaction
–Cl	Cu(I)Cl	may be a radical reaction
–Br	Cu(I)Br	may be a radical reaction
–I	KI	best way to add iodine
–Ar	ArH	Friedel-Crafts arylation
–H	H_3PO_2 or EtOH/H$^+$	reduction of ArN_2^+

ortho- and para- Product Mixtures

We used the nitration of toluene to give both the *para*-nitro **4** and the *ortho*-nitro compounds **48**. In fact the reaction gives a mixture. This is acceptable providing the compounds can be separated and especially so if industry does the job on a very large scale, as here. The synthesis of the sweetener saccharine is a good example. Saccharine **50** is a cyclic imide: that is a double amide from one nitrogen atom and two acids. If we disconnect the C–N and S–N bonds the two acids—one carboxylic and one sulfonic—are revealed **51**. Both groups are *meta*-directing so we must do FGI to convert one of them into an *ortho,para*-directing group and we can use the same oxidation reaction we met at the start of the chapter (**4** to **3**). Now **52** can be made by sulfonation.

In practice chloro-sulfonic acid is used as this gives the sulfonyl chloride directly. You may be surprised at this, thinking that Cl might be the best leaving group. But there is no Lewis acid here. Instead the very strong chloro-sulfonic acid protonates itself to provide a molecule of water as leaving group (see workbook).

The reaction gives a mixture of the *ortho-* **53** and *para-* **54** products. The *ortho*-compound is converted into saccharine by reaction with ammonia and oxidation and the *para*-compound toluene-*p*-sulfonyl chloride **54**, or tosyl chloride, is sold as a reagent for converting alcohols into leaving groups.

References

1. *Drug Synthesis*, vol. 1, page 9; H. Salkowski, *Ber.*, 1895, **28**, 1917.

2. P. H. Gore in *Friedel-Crafts and Related Reactions*, ed. G. A. Olah, Vol III, Part 1, Interscience, New York, 1964, p. 180.

3. W. Weinrich, *Ind. Eng. Chem.*, 1943, **35**, 264; S. H. Patinkin and B. S. Friedman in ref. 2, vol. II, part 1, p. 81.

4. E. L. Martin, *Org. React.*, 1942, **1**, 155.

5. *Pesticides*, p. 154; *Pesticide Manual*, p. 537.

6. Saunders, *Top Drugs*, Oxford University Press, p. 44.

7. H. T. Clarke and R. R. Read, *Org. Synth. Coll.*, 1932, **1**, 514.

3 Strategy I: The Order of Events

Background Needed for this Chapter Reference to Clayden, *Organic Chemistry:*
Electrophilic aromatic substitution; chapter 22. (Electrophilic Aromatic Substitution)

Alternating with instructional chapters, like the last one, will be strategy chapters, like this one, which discuss reasons for choosing one route rather than another: in other words the overall plan rather than the individual steps. In this chapter we shall examine the order of events, using the synthesis of aromatic compounds as examples. The details are specific but the guidelines general.

Guideline 1: Consider the effects of each functional group on the others. Add first (that is disconnect last) the one that will increase reactivity in a helpful way. So, for aromatic compounds, introduce first that group that helps, by reactivity or direction, the introduction of the others.

The analysis of the perfumery compound **1** could be tackled by two possible first disconnections. Friedel-Crafts alkylation **a** would work reasonably well with the secondary alkyl halide **2** but the ketone in **3** is *meta*-directing and would give the wrong product. Friedel-Crafts acylation **b** would give the right product as the alkyl group in **4** is *ortho,para*-directing. Further the alkyl group in **4** is activating while the ketone in **3** is deactivating.

Analysis

The synthesis[1] is straightforward providing we alkylate first and acylate second. The branched alkyl group in **4** ensures that the *para*-ketone **1** will be the main product by steric hindrance.

Synthesis

Organic Synthesis: The Disconnection Approach. Second Edition   Stuart Warren and Paul Wyatt
© 2008 John Wiley & Sons, Ltd

Our next example emphasises one aspect of guideline 1. Some functional groups are so deactivating that it is difficult to do any further chemistry once they have been inserted. In other contexts it may be that they are so unstable that we would not wish to risk any further reactions. Musk ambrette **6** is a synthetic musk, essential in perfumes to enhance and retain the fragrance. It has five substituents round a benzene ring. Two of these, the nitro groups, are so deactivating that we want to add them last. So we disconnect them first.

Next we could disconnect either the methyl group **7a** or the *t*-butyl group **7b** with Friedel-Crafts alkylation in mind. There are many reasons for choosing **7b**. The MeO group and the alkyl group in **8** and **9** are *ortho,para*-directing but the MeO group wins because it has a lone pair on oxygen that is delocalised into the aromatic system. So **9** will give the wrong product but **8** will give the right product. In addition, alkylation works well with a *t*-butyl group (S_N1 mechanism) but badly with a methyl group. Our starting material **8** is the methyl ether of available *meta*-cresol **10** so is easily made by methylation of the phenol. Only experience would show that alkylation of **8** puts the *t*-butyl group *ortho* and not *para* to the MeO group.[2]

Guideline 2: Changing one functional group into another may alter reactivity dramatically. Changing an alcohol or phenol to a *t*-Bu ether increases steric hindrance. Alcohols and aldehydes or ketones are easily interconverted by redox reactions. The carbonyl compounds are electron-withdrawing, alcohols weakly electron-donating. Most dramatically for aromatic compounds, the nitro group is powerfully electron-withdrawing, deactivating and *meta*-directing while the amino group, often made by reduction of a nitro group, is strongly electron-donating, activating and *ortho,para*-directing. In an analysis featuring FGI, it may pay to consider at which stage to carry out some other reaction.

A simple example is the tetrachlorocompound **11** clearly made from toluene by some form(s) of chlorination. In fact we must change the *meta*-directing CCl_3 group into a *ortho,para*-directing methyl group before we disconnect the Ar-Cl bond.

This compound **11** was actually used[3] to make the trifluorocompound **13**. The chlorination of toluene with Lewis acid catalysis gives mostly **12** and chlorine and PCl_5 does the, probably radical, chlorination of the methyl group.

Guideline 3: Some substituents are difficult to add so it is best to start with them already present. It is not necessary to start all syntheses of aromatic compounds from benzene: a glance at any supplier's catalogue will show the great range of aromatic compounds available. The chief examples of such substituents are phenols and the ethers derived from them as there is no simple reagent for electrophilic oxygen. But a methyl group and primary alkyl groups in general, are also difficult to add as Friedel-Crafts alkylation with primary alkyl halides leads to rearranged products.[4]

The trisubstituted benzene **14** was used by Woodward as a starting material for his synthesis of the natural product reserpine.[5] It too has to be made. We shall not add the MeO group but buy anisole (methoxybenzene) as starting material. Both nitrogens will be added by nitration but in which order?

The MeO group is *ortho,para*-directing so nitration of anisole **15** will give mostly the *para* product **16** (steric hindrance). Nitric acid alone is needed: a mixture of nitric and sulfuric acids gives 2,4-dinitroanisole as the MeO group is activating.[6] This also shows that a second nitrogen cannot be introduced in the right place from **16**. However, reduction to the amine **17**—many reagents could be used but Ti(III) gives good results[7]—gives a more powerful activating group that might direct nitration to the right position (Guideline 2).

In practice, there is too much activation in **17** and attempted nitration oxidised the molecule. The amine must be acetylated first and then, without isolation of **18**, can be nitrated (with nitric acid alone) to give **19**. Hydrolysis of the amide gives **14** in excellent yield.[8]

Guideline 4: Some disubstituted compounds are also readily available and they may contain a relationship (especially *ortho*) that is difficult to achieve by electrophilic substitution. Here is a selection: a supplier's catalogue will reveal more.

salicylic acid
(also aldehyde)

anthranilic acid

phthalic
anhydride

mesitylene

ortho (catechol)
meta (resorcinol)
para (quinol)

ortho, meta and
para cresols

diphenyl

A good example is **21** needed for the synthesis of the GSK anti-asthma drug salbutamol **20**. This ketone **21** could be made by a Friedel-Crafts acylation of **22**, which turns out to be salicylic acid, with acetyl chloride.

20; salbutamol

21

22; salicylic acid

acetyl
chloride

This synthesis is easier than it may seem as the free phenol, rather then interfering, can be acylated to give the ester **23** which rearranges with $AlCl_3$ to give **21** directly.[9] Even the intermediate **23** is available and cheap–it is aspirin.

22; salicylic acid

23; aspirin

TM21

Guideline 5: Some groups can be added to the ring by nucleophilic substitution. This is mechanistically more difficult than electrophilic substitution and requires an electron-withdrawing activating group such as nitro or carbonyl *ortho* or *para* to a normal leaving group such as a halide (chapter 2). Fortunately nitration or Friedel-Crafts acylation of halocompounds puts the activating group in the right position for nucleophilic substitution. So Friedel-Crafts acylation of fluorobenzene **24** gives the ketone **25** and displacement of fluoride by the addition–elimination mechanism[10] gives the amine **26**.

24

25

26

If this kind of activation is not available, nitrogen can be displaced from diazonium salts by the S_N1 mechanism. The acid **27** was needed at Hull University in work on liquid crystals.[11]

The skeleton is diphenyl (Guideline 4) which reacts in the *para*-positions with electrophiles. The chlorination is difficult therefore and we need to replace the CO_2H group with a group more electron-donating than the phenyl ring. An amine is the answer **28** and that soon takes us back to diphenyl.

The nucleophile to introduce the CO_2H group is cyanide ion, used as its Cu(I) salt, and the amine in **29** must be acylated to prevent over-chlorination (compare **18**).

Guideline 6: If a series of reactions must be carried out, start with one that gives a single product unambiguously and not one that would give a mixture. With aromatic compounds if you need to add both *ortho* and *para* substituents, putting in the *para* substituent first may be less ambiguous than the reverse.

Compound **33** was needed to make some antimalarial drugs.[12] We prefer not to disconnect the OEt group (Guideline 3) and there are good reactions—nitration and chloromethylation—that would go in the right position (*ortho* or *para*) to the activating OEt group. Either disconnection **a** or **b** could be tried first.

We expect the *para* product to be the major product from either reaction on ether **36** (steric hindrance) so it makes sense to nitrate first. There is also a danger that nitration of **34** might oxidise the CH_2Cl group to CHO or even CO_2H. The synthesis works well if nitration is carried out first.[12]

It should be obvious that not all of these six guidelines will be relevant in every synthesis–indeed some may even contradict others. It is a matter for judgement and then laboratory trial to select a good route. As always, several different strategies may be successful.

References

1. G. Baddeley, G. Holt and W. Pickles, *J. Chem. Soc.*, 1952, 4162.
2. M. S. Carpenter, W. M. Easter and T. F. Wood, *J. Org. Chem.*, 1951, **16**, 586.
3. H. S. Booth, H. M. Elsey and P. E. Burchfield, *J. Am. Chem. Soc.*, 1935, **57**, 2066.
4. Clayden, *Organic Chemistry*, page 573.
5. R. B. Woodward, F. E. Bader, H. Bichel, A. J. Frey and R. W. Kierstead, *Tetrahedron*, 1958, **2**, 1.
6. *Vogel*, page 1256.
7. T.-L. Ho and C. M. Wong, *Synthesis*, 1974, 45.
8. P. E. Fanta and D. S. Tarbell, *Org. Synth.*, 1945, **25**, 78.
9. D. T. Collin, D. Hartley, D. Jack, L. H. C. Lunts, J. C. Press, A. C. Ritchie and P. Toon, *J. Med. Chem.*, 1970, **13**, 674.
10. Clayden, *Organic Chemistry*, chapter 23.
11. D. J. Byron, G. W. Gray, A. Ibbotson and B. M. Worral, *J. Chem. Soc.*, 1963, 2253;   D. J. Byron, G. W. Gray, and R. C. Wilson, *J. Chem. Soc. (C)*, 1966, 840.
12. J. H. Burckhalter, F. H. Tendick, E. M. Jones, P. A. Jones, W. F. Holcomb and A. L. Rawlins, *J. Am. Chem. Soc.*, 1948, **70**, 1363.

4 One-Group C–X Disconnections

Background Needed for this Chapter References to Clayden, *Organic Chemistry:*
Chapter 12: Nucleophilic Substitution at the Carbonyl (C=O) Group.
Chapter 17: Nucleophilic Substitution at Saturated Carbon.

We started with aromatic compounds in chapters 2 and 3 because the position of disconnection needed no decision. We continue with ethers, amides and the like because the position of disconnection is again easily decided: we disconnect a bond joining the heteroatom (X) to the rest of the molecule: a C–O, C–N or C–S disconnection. We call this a 'one-group' C–X disconnection because we need to recognise only one functional group (ester, ether, amide etc.) to know that we can make the disconnection.

The corresponding reactions are mostly ionic involving nucleophilic displacement by S_N1, S_N2 or carbonyl substitution with amines, alcohols and thiols on carbon electrophiles. The normal polarity of the disconnection **1** will be a cationic carbon synthon **2** and an anionic heteroatom synthon **3** represented by acyl or alkyl halides **4** as electrophiles and amines, alcohol or thiols **5** as nucleophiles.

It is possible to use the reverse polarity with a nucleophilic carbon synthon **6** and an electrophilic heteroatom synthon **7** but only with second or third row elements such as S, Si, P and Se. These synthons are represented by organometallic compounds **8** or **9** and compounds **10** such as RSCl, Me_3SiCl and Ph_2PCl and we shall consider these later.

Carbonyl Derivatives RCO.X

We start with acid derivatives since we almost always choose to disconnect the bond between the heteroatom and the carbonyl group. So we make esters **11** and amides **13** from acid (acyl) chlorides **12** and alcohols or amines.

Organic Synthesis: The Disconnection Approach. Second Edition   Stuart Warren and Paul Wyatt
© 2008 John Wiley & Sons, Ltd

11; ester   alcohol   12   amine   13; amide

The ester **14**, used both as an insect repellent and as a solvent for perfumery, is easily made this way. The analysis reveals two available compounds: benzyl alcohol **15** and benzoyl chloride **16**. Combining the two with pyridine as solvent and catalyst gives the ester **14**.

14   15   16

Acid chlorides are often used in these syntheses because they are the most electrophilic of all acid derivatives and because they can be made from the acids themselves with PCl_5 or $SOCl_2$. The other important acid derivatives can all be made from acid chlorides or from any compound above them in the chart of reactivity. So you can make amides from acid chlorides, anhydrides or esters but it is very difficult to make any other derivatives from amides. All derivatives except amides can easily be made from the acids themselves.

A simple example is the weedkiller propanil **17** used on rice fields. Amide disconnection gives the amine **18** obviously made from *o*-dichlorobenzene **20** by nitration and reduction. All positions around the ring in **20** are about the same electronically but steric hindrance will lead to **19** being the major product.

17; propanil   18   19   20

The synthesis is very simple.[1] The only point worth noting is the use of catalytic hydrogenation for the reduction rather than the very messy tin and HCl. Industry greatly prefers catalytic methods with no toxic by-products.

The alternative disconnection, between the alkyl group and the heteroatom **21**, is also acceptable but it would require an S_N2 reaction between the anion of the acid **22** and an alkyl halide. Reactions at carbonyl groups are much more reliable than S_N2 reactions and are usually preferred. Only if the S_N2 reaction is exceptionally good, as it would be to make the ester **23**, is this route preferred.

The Synthesis of Ethers

This same question of which bond to disconnect can be much more significant in the synthesis of ethers. With many ethers, like the gardenia perfume compound **24**, it doesn't matter much. The starting materials will be an alcohol **26** or **27** and an alkyl halide, say **25** or **28**.

The reaction will be carried out by treating the alcohol with a base strong enough to form the anion—sodium hydride is a favourite as the hydride ion (H^-) is extremely hard and acts only as a base, never as a nucleophile. Either alcohol is available, either would give a nucleophilic anion. Either chloride is available, both react well in S_N2 reactions. We prefer route **a** as benzyl chloride **25** is more reactive and cannot undergo elimination while **28** just might.[2]

Choosing a Route Mechanistically

In other cases, the choice can be made because only one of the two S_N2 reactions will work well. The allyl phenyl ether **32** can be disconnected to bromobenzene **30** and the allylic alcohol **31** or to phenol **33** and the allylic bromide **34**.

We might prefer route **b** because phenol is much more acidic (pK_a 10) than an alcohol such as **31** (pK_a about 15) and so a weaker base such as NaOH can be used. But the main reason to prefer route **b** is that route **a** will not work. Nucleophilic substitutions on bromobenzene do not work[3] while those on allylic halides such as **34** work very well indeed[4] (chapters 2 and 3).

Notice that a disconnection we did not consider for a moment would come from reversal of polarity (in either C–O bond). Trying to improve disconnection **a**, we might suggest a nucleophilic phenyl species such as **36**—perfectly all right—and an electrophilic oxygen reagent such as the hypochlorite **37**—definitely not all right. If **37** existed it would be dangerously explosive.

Since alkyl halides are made from alcohols by treatment with reagents such as PBr$_3$ or HCl and a Lewis acid, it may make sense in designing the synthesis of ethers to write the two alcohols as starting materials and decide later which to convert into an electrophile. For ether **24** the two alcohols would be **27** and **26**.

Either could be converted into the corresponding chloride or bromide. Benzyl chloride **25** is easy to make as both S$_N$1 and S$_N$2 reactions work well at a benzyl group. Formation of the primary alkyl bromide **38** needs more vigorous conditions but gives 83% yield.[5] In practice either compound can be bought cheaply.

If the ether is symmetrical, ROR, it is enough to treat the alcohol ROH with acid. Note that this applies only to symmetrical ethers. You would not get a good yield if you treated the mixture of **27** and **26** with acid: as well as the cross product, each alcohol would dimerise and there would be three products that would be very difficult to separate. This sort of question is considered in the next chapter.

The Synthesis of Sulfides

Unsymmetrical sulfides **39** need the same disconnection we have just used for ethers. The anion **41** of a thiol **42** will combine with an alkyl halide **40** to make a new C–S bond. The reaction is much easier with sulfur. Thiols are more acidic than alcohols, just as H_2S is more acidic than water. Sulfide anions **41** are more nucleophilic towards saturated carbon than are alkoxides and the risk of elimination is much less.

The acaricide (kills mice and ticks) chlorbenside **43** is disconnected to give an acidic thiophenol **44** and a reactive alkyl halide **25**. The synthesis merely combines these two in ethanol with NaOEt as base.[6]

Symmetrical sulfides can be made from the alkyl halide and Na_2S as the product from the first step is the monoanion needed to make the second C–S bond. The synthesis[7] is just to combine the alkyl bromide with Na_2S in ethanol: dipropyl sulfide (R = Et) is formed in 91% yield from the bromide while dibenzyl sulfide (R = Ph) is made in 83% yield from benzyl chloride **25**.

Summary of Compounds Made from Alcohols

Many nucleophiles we haven't mentioned can be used in these reactions. In every case a nucleophilic heteroatom displaces a leaving group from a compound derived from an alcohol.

We have focussed on alkyl halides but tosylates from TsCl and mesylates from MsCl can be used too. The conversion of alcohols to chlorides and bromides is discussed earlier in this chapter and the combination of reagents used to make thiols is discussed in the next chapter.

TsCl =

toluene-*p*-sulfonyl chloride

MsCl =

methane sulfonyl chloride

References

1. *Pesticides*, p. 152; *Pesticide Manual*, p 446; W. Schäfer, L, Eue and P. Wegler, *Ger. Pat.*, 1958, 1,039,779; *Chem. Abstr.*, 1960, **54**, 20060i.

2. *Perfumes*, page 226; G. Errera, *Gazz. Chim. Ital.*, 1887, **17**, 197.

3. Clayden, *Organic Chemistry*, chapter 23.

4. Clayden, *Organic Chemistry*, chapter 17.

5. *Vogel*, page 562.

6. J. E. Cranham, D. J. Higgons and H. A. Stevenson, *Chem. and Ind. (London)*, 1953, 1206;   H. A. Stevenson, R. F. Brookes, D. J. Higgons and J. E. Cranham, *Brit. Pat.*, 1955, 738,170; *Chem. Abstr.*, 1956, **50**, 10334b.

7. *Vogel*, page 790.

5 Strategy II: Chemoselectivity

Background Needed for this Chapter References to Clayden, *Organic Chemistry:*
Chapter 8: Acidity, Basicity, and pK_a.
Chapter 24: Chemoselectivity: Selective Reactions and Protection.

If a molecule has two reactive groups and we want to react one of them and not the other we need chemoselectivity. Under this heading we can consider:

1. The relative reactivity of two different functional groups, such as NH_2 and OH.

2. The reaction of one of two identical groups: we might want to make the *mono*ether **5**.

3. The reaction of a group once when it might react twice as in thiol synthesis.

Guideline 1: If two groups have *unequal* reactivity, the *more* reactive can be made to react alone.

The amide **2** is paracetamol, the popular analgesic. Amines are much more nucleophilic than phenols (compare the basicities of ammonia and water) so reaction with acetic anhydride gives

Organic Synthesis: The Disconnection Approach. Second Edition   Stuart Warren and Paul Wyatt
© 2008 John Wiley & Sons, Ltd

the amide we want without any of the ester **3**. The aminophenol **1** can be made by methods explained in chapter 2.

The synthesis is straightforward. Nitration of phenol needs only dilute nitric acid and the reduction is best carried out catalytically.[1]

The Synthesis of Cyclomethycaine

Selectivity between two oxygen nucleophiles might sound more difficult but when one is an alcohol and the other a carboxylic acid, there is no problem. The local anaesthetic cyclomethycaine **12** is obviously made from the carboxylic acid **13** and an amino-alcohol. The acid **13** is our concern. We disconnect the ether linkage **13a** on the alkyl side so that the S_N2 reaction works. Chemoselectivity now arises as **15** has OH and CO_2H functional groups. Which will act as a nucleophile?

The answer is that it depends on the pH. Below about pH 5 **15** is a neutral compound. The OH group is more nucleophilic than the delocalised CO_2H group but is not nucleophilic enough to react with an alkyl halide. We can increase its reactivity by adding base and, between about pHs 5 and 10, it exists as the carboxylate anion **16**. We don't want this as it will react at CO_2^- rather than at OH. But at pHs above about 10 it exists as the dianion **17** and now at last ArO^- is more nucleophilic than CO_2^-.

All we need to do is to use two equivalents of a strong enough base with a suitable alkyl halide and we shall make our ether. As **14** is a rather unreactive secondary alkyl halide, we need a good leaving group such as iodide.[2]

Guideline 2: If one functional group can react twice, the product of the first reaction will compete with the reagent. The reaction will stop cleanly after one reaction only if the starting material is more reactive than the product.

So the reaction of an alkyl halide with NaSH or Na$_2$S cannot usually be made to stop after one alkylation as the anion of the first product is at least as nucleophilic as HS$^-$ or S^{2-}. This is obvious in reactions with Na$_2$S. Less obviously with NaSH the first reaction **18** gives the thiol **19** but this is in equilibrium with RS$^-$ and a second displacement **20** gives the sulfide **21**. We shall see shortly how to get round this problem. A more important example—the failure of the alkylation of ammonia to give a useful amine synthesis—has chapter 8 to itself.

But some reactions of this sort are successful. The synthesis of chloroformates from alcohols and phosgene (or a phosgene equivalent) is a useful example. We shall need benzyl chloroformate **21** in the next section to introduce an important protecting group. As it is an ester, disconnection to benzyl alcohol **22** and phosgene **23** looks good. But the product **21** is itself also an acid chloride and looks as though it might react again to form dibenzyl carbonate. But in this case there is delocalisation **24a** in the product that is not present in phosgene and the carbonyl group of **21** is much less electrophilic than that of phosgene. The synthesis is successful.[3] The halogenation of ketones in acid solution (chapter 7) is another example where a reaction occurs only once.

Guideline 3: Problems from guidelines 1 and 2 may be solved by protecting groups.

If we want to react the *less* reactive of two functional groups we protect the *more* reactive. If we want a reagent to react *once* when it could react *twice* we protect the reagent. A protecting group is something added to a functional group that reduces or eliminates unwanted reactivity. It must be easy to add and easy to remove as we are adding two steps to our synthesis. In an ideal world, no protecting groups would be needed but just look at any recently published synthesis of any moderately complex molecule and you will see several protecting groups. Protecting groups have chapter 9 to themselves.

A classic case is amino acid chemistry. The amine is more nucleophilic than the carboxyl group so, if we want to use the carboxyl group as a nucleophile, we must protect the amino group. Benzyl chloroformate **22** is often used in this way. It cleanly acylates the amino group to give the carbamate **26** whose nitrogen atom is much less nucleophilic because of further conjugation **27**. If you compare **23, 22,** and **26** you will see the carbonyl group becoming less and less electrophilic. The anion of **26** will now react at oxygen (CO_2^- in basic solution) with electrophiles.

22   25   26   27

The synthesis of thiols **19** is well managed if thiourea **28** is used instead of NaSH or Na_2S as the nucleophile. Thiourea is itself highly delocalised but it is still a good nucleophile for saturated carbon through the sulfur atom **29**. The product is a thiouronium salt **30** and is not nucleophilic at all, being a cation. Hydrolysis with aqueous base liberates urea **31** and the thiol **19**.

28; thiourea   29   30   19   31; urea

The sedative and tranquilliser captodiamine **32** contains two sulfides and four C–S bonds. Disconnection next to the more central sulfur atom **32** could be of either C–S bond. The one chosen leads to an available (but unpleasant) amine **34** and a thiol **33**.

32; captodiamine   33   34

Thiol **33** was made by the thiourea method from **35** and further disconnection by the methods from previous chapters takes us back to available benzene thiol **39**.

33   35   36   37

37a   38   39

The first sulfide needs only a weak base as benzene thiol is acidic and the electron-donating BuS group directs *para*. The rest is straightforward.[4]

PhSH —BuCl/Na_2CO_3→ PhSBu —PhCOCl/$AlCl_3$→ 37 —1. $NaBH_4$ 2. $SOCl_2$→ 35 —1. $(H_2N)_2CS$ 2. NaOH H_2O→ 33 —34/base→ TM32

39   38

Guideline 4: One of two identical groups may give a reasonable yield by a number of methods.

(a) In the same way as guideline 2, if the product from the first reaction is less reactive than the starting material, double reaction may be avoided.

Partial reduction of *meta*-dinitrobenzene is an example. Nitration of nitrobenzene **40** is difficult but succeeds with fuming nitric acid (about 90% HNO_3) in sulfuric acid and gives only the *meta* product **41** as expected from the deactivating nitro group.[5] Reduction with NaHS cleanly gives *m*-nitroaniline by reduction of just one nitro group.[6] The reason is that reduction is electron donation and the very electron-withdrawing nitro group encourages this. So the product **42** is less easily reduced than the starting material **41**.

(b) If the starting material and the mono-reacted product are of roughly equal reactivity, reaction with one equivalent of reagent may give a moderate yield by the statistical method.

Treatment of the symmetrical diol **43** with strong base gives a mixture of monoanion **44**, starting material **43** and the dianion. If the mixture were perfectly statistical, and each species equally nucleophilic, we should expect 50% **44** and 25% each of **43** and the dianion. It turns out that the starting material does not react with EtBr and there is evidently less of the dianion, presumably because the two anions are near enough to destabilise each other. The yield of the monoether **45**, used in a synthesis[7] of vitamin E, is 65%, substantially higher than 50%.

(c) The best method is to combine the two identical functional groups into one functional group that reacts once to give a product of much lower reactivity than the starting material. This cannot always be arranged but cyclic anhydrides such as **48** are very useful reagents. If either **47** or **48** are combined with methanol in acidic solution, the diester **46** is formed. But in basic solution the anion of the monoester **49** is the product and the remaining acid group can be transformed into, say, the acid chloride **50** so that each carboxylic acid has been activated in a different way.

The method works because methoxide attacks either carbonyl group **51** displacing the carboxylate anion. The product under the reaction conditions is **52** and methoxide will not attack another anion to make the diester. Neutralisation gives the half ester[8] **49**.

The cyclic anhydride can also be used in Friedel-Crafts reactions. Note the position of acylation on **53**–*para* to the chlorine and not to the methyl group–and that acylation has occurred only once. The product **54** was used in the synthesis of fungicidal compounds.[9]

These methods, particularly (a) and (b), depend on efficient separation of the starting material and the products from mono- and di-reaction. If separation cannot be achieved then a very good method is needed.

A Warning

If two groups are nearly but not quite identical, do not attempt to react one and not the other. Examples include the diols **55** and **56** where one extra methyl group is not enough to make a significant difference in reactivity. Attempts to make a specific monoether from either are doomed.

References

1. L. Spiegler, *U. S. Pat.*, 1960, 2,947,781; *Chem. Abstr.*, 1961, **55**, 7353f; M. Freifelder, *J. Org. Chem.*, 1962, **27**, 1092.
2. S. P. McElvain and T. P. Carney, *J. Am. Chem. Soc.*, 1946, **68**, 2592.
3. M. Bergmann and L. Zervas, *Ber.*, 1932, **65**, 1192.
4. *Drug Synthesis* page 44; O. H. Hubner and P. V. Petersen, *U. S. Pat.*, 1958, 2,830,088; *Chem. Abstr.*, 1958, **52**, 14690i.
5. *Vogel*, page 855.
6. *Vogel*, page 895; H. H. Hodgson and E. R. Ward, *J. Chem. Soc.*, 1949, 1316.
7. L. I. Smith and J. A. Sprung, *J. Am. Chem. Soc.*, 1943, **65**, 1276.
8. P. Ruggli and A. Maeder, *Helv. Chim. Acta*, 1942, **25**, 936.
9. T. Tojima, H. Takeshiba and T. Kinoto, *Bull. Chem. Soc. Jpn.*, 1979, **52**, 2441.

6 Two-Group C–X Disconnections

Background Needed for this Chapter References to Clayden, *Organic Chemistry:*
Chapter 10: Conjugate Addition.
Chapter 21: Formation and Reaction of Enols and Enolates.

One-Group and Two-Group C–X Disconnections

Asked to make the sulfide **1** you would not hesitate to disconnect a C–S bond, choosing the one between the sulfur and the aliphatic part of the molecule to ensure a good S_N2 reaction. There is only one functional group in the target molecule **1** so you have to choose a one-group C–X disconnection.

If you were asked to make the sulfide **4** you might very reasonably take the same decisions, proposing the same sulfur compound **2** as nucleophile and the alkyl bromide **5** as electrophile.

There is nothing wrong with this suggestion except that it ignores the other functional group—the ketone—in the target molecule **4** and so misses an opportunity for a two-group disconnection. Our message in this chapter is going to be that two-group disconnections are better than one-group disconnections. Reverting to synthons for a moment, the sulfur synthon **2** is the same as the reagent but the carbon synthon **6** might make you think of a different reagent.

Organic Synthesis: The Disconnection Approach. Second Edition   Stuart Warren and Paul Wyatt
© 2008 John Wiley & Sons, Ltd

The idea with two-group disconnections is that we recruit the other functional group to help us discover a better reagent. Here the carbonyl group can make the cationic centre in **6** electrophilic if we simply add a double bond to the structure.

The reaction is conjugate addition of the thiolate anion **2** to the enone **7** making an enolate intermediate that captures a proton from PhSH **8** to give the target molecule **4** and regenerate the nucleophile **2**.

This route using **7** is better than the first suggestion using **5** for several reasons.

1. There is no need to waste an atom of bromine to provide a leaving group: the enone **7** is naturally electrophilic at the right carbon atom.
2. The two functional groups in the target molecule co-operate in making the new C–S bond.
3. No strong acids, bases or high temperatures are needed as the enolate intermediate regenerates the reagent **2** so only catalytic weak base is needed.
4. The bromide **5** is likely to eliminate under the reaction conditions to give **7** anyway.

Recognising a Two-Group C–X Disconnection

The key step is recognising the *relationship* between the two functional groups. To do this, number the *carbon atoms* bearing the functional groups. It doesn't matter which you call 1—only the relationship matters. Here we see **4a** that we have a **1,3-diX** relationship. That means that the two functionalised carbon atoms have a 1,3-relationship. Knowing that, we can choose conjugate addition as our reaction and do the disconnection **4b** we have already done and reveal **2** and **7** as our reagents.

To start with you may like to draw the synthons and, by inspecting the carbon synthon, decide which electrophilic reagent to use. But as this chapter develops, you will see that there is a particular chemistry used to make each different relationship (e.g. a 1,3-relationship suggests conjugate addition) and you may soon not bother with the synthons but write the reagents directly. This is a matter for personal choice.

The 1,3-diX Relationship

Since we are using conjugate addition, it is essential to have an electron-withdrawing group, usually a carbonyl group but it could be CN for example, in the right position. The disconnection in general terms is this:

The nucleophilic reagent will depend on the heteroatom. If X=O or S, base will probably be necessary, but if X=N, the amine itself should be nucleophilic enough to do conjugate addition. An example would be the amino ester **13**. Numbering the carbon atoms **13a** reveals the 1,3-relationship and C–N disconnection gives the secondary amine **14** and ethyl acrylate as reagents.

This is the time to reveal a potential problem. In this synthesis we want conjugate addition. But we might on another occasion want to make the amide **18** so how do we control whether the nucleophile adds direct to the carbonyl group or by conjugate addition **17**? In general the reactivity of the electrophile is crucial. Very electrophilic compounds such as acid chlorides or aldehydes tend to prefer direct addition while less electrophilic compounds such as esters or ketones tend to do conjugate addition.

What if There Is No Carbonyl Group?

The amino alcohol **19** has a 1,3-diX relationship but no carbonyl group. So we introduce one by FGI. We could have an ester or an aldehyde. An aldehyde would be easier to reduce but there is a danger of direct addition. So we choose an ester (it doesn't matter which).

The synthesis is straightforward and we shall need $LiAlH_4$ to reduce the ester.

Supposing there is no oxygen-based functionality at all, as in the diamine **23**? It is not necessary to have a carbonyl group for conjugate addition, in fact a nitrile is much better. So we do FGI on the primary amine and disconnect the secondary amine Et$_2$NH from acrylonitrile **25**. The synthesis is to mix those two and reduce **24** catalytically or with LiAlH$_4$.

Examples of 1,3-Difunctionalised Compounds

Enantiomerically pure chloro-ester **26** was needed for an investigation into the stereochemistry of the Friedel-Crafts reaction. Disconnecting the ester we reach the one piece of carbon skeleton and see that it has a 1,3-diX relationship **27**. However we need a carbonyl group and an ester **28** should ensure conjugate addition of chloride to **29**.

The acid itself was chosen for the conjugate addition as the intermediate can then be resolved by crystallisation of the quinine salt. Conjugate addition of HCl was successful and the acid was reduced to the alcohol with LiAlH$_4$ before esterification in the usual way.[1]

The aminoether **30** containing a seven-membered ring has a 1,3-relationship but no carbonyl group. We could remove the seven-membered ring and put a carbonyl group at C-3 but a shorter synthesis comes from the nitrile **31** as we can add the alcohol **32** in one piece to acrylonitrile **25**.

The synthesis is a simple two-stage process with catalytic hydrogenation used for reduction of the nitrile.[2]

The 1,2-diX Relationship

The 1,2-diX relationship presents a different series of opportunities in which we use the second functionality to make the right carbon atom electrophilic. The amino, thio- and alkoxy- alcohols **33** to **35** all fit the pattern **36** and can be disconnected to the usual heteroatom nucleophile and the synthon **37**.

If you don't see at once what reagent will be used for the synthon **37**, you are not alone. How can we use the other OH group at C-1 to make C-2 electrophilic? One way to visualise the answer is to imagine what would happen if you actually made the cation **37**. It would instantly cyclise **38** to form a three-membered ring **39** that could lose a proton to give the epoxide **40**. Epoxides are strained ethers and react with nucleophiles such as amines **41** to give **42** and hence the aminoalcohol **33**.

It doesn't matter which end of the symmetrical epoxide is attacked by the nucleophile—the same product **42** is formed. If there is a substituent at either end of the molecule **43** or **45** we can still make the 1,2-diX disconnection but the 'two' epoxides **44** are the same. This is clearly a problem. In fact the nucleophile will prefer to attack the less substituted end of the three-membered ring **46** so we can make **45** but not **43** this way.

The disconnection **47** gives a different synthon **48** at the carbonyl oxidation level and the best reagents are the α-halo carbonyl compounds **49**. At first this looks like a one-group disconnection but there are two reasons why it isn't. The presence of the carbonyl group makes the S$_N$2 displacement of halide enormously faster and the α-halo carbonyl compounds **49** are made from the ketone **51** by acid-catalysed halogenation of the enol **50**.

Examples of 1,2-Difunctionalised Compounds

The ether **52** was needed to study the Claisen rearrangement and can be disconnected in 1,2-diX fashion as the epoxide **54** will be attacked at the less hindered end by the anion of **53**.

The synthesis involves treating the allylic alcohol **53** with NaH to make the anion **55** and combining that with the epoxide **54** easily made from styrene.[3]

Compounds like the ester **56** were briefly mentioned in chapter 4 and we can now show how they can be made by a two-group disconnection. The electrophile will be the α-halo carbonyl compound **57** and the nucleophile the anion of a carboxylic acid.

The bromide **57** is made by direct bromination of the ketone **59** and only the very weak base NaHCO$_3$ is needed to make the anion of the carboxylic acid. This reaction shows just how electrophilic such α-halo carbonyl compounds must be as carboxylate anions are very weak nucleophiles. Compounds **56** are therefore derivatives of carboxylic acids. They are highly crystalline and can be used to characterise and purify such acids.[4]

The Pfizer anti-fungal compound fluconazole **60** is a more advanced example of such disconnections.[5] It has two identical 1,2-diX relationships between nitrogen and the OH group. You might think that we can make both the same way, but not so. The first disconnection is easy: we want the aromatic amine triazole **62** to combine with the epoxide **61** at its less substituted end.

But how are we to make the epoxide **61**? The obvious route is by epoxidation of the alkene **63**. The alkene **63** could be made by a Wittig reaction (chapter 15) on the ketone **64** or directly by sulfur ylid chemistry (chapter 30).

The ketone **64** still has a 1,2-diX relationship but at the carbonyl oxidation level, so we disconnect to another molecule of triazole **62** and the α-halo ketone **66** is easily made by a Friedel-Crafts reaction using available chloroacetyl chloride. This time we buy the 1,2-diX relationship in the form of chloroacetyl chloride.

The synthesis follows the analysis exactly giving fluconazole **60** in only 5 steps from available starting materials. This synthesis should demonstrate that important modern drugs are made by the style of reactions that you are meeting in this book.[6]

The 1,1-diX Relationship

The label '1,1-diX' may look strange but all it means is that the two functional groups are joined to the same carbon atom. You already know how to make acetals **68**: you combine an aldehyde **67** with an alcohol, say methanol, in acidic solution. The disconnection **68a** is therefore of both C–O bonds. This reveals a valuable truth: two heteroatoms joined to the same carbon atom are at the carbonyl oxidation level (two C–O bonds to the same C atom in both **67** and **68**) and the TM is probably made from a carbonyl compound.

Again you may think that we are not using the two groups in cooperation. But we are. The key step in acetal formation and hydrolysis is the expulsion of one OR group by the other. In the synthesis, protonation of the hemiacetal **69** is followed by expulsion of a molecule of water **70** and the addition of the second molecule of methanol **71** is possible because of the first. In the hydrolysis, these steps happen in reverse. Acetals are not 'just ethers'—they are more reactive compounds because of the two RO groups joined to the same carbon atom.

If the two heteroatoms are the same, it is usually best to disconnect both C–X bonds, choosing the ones to the same carbon atom, and write a carbonyl group at that atom. The heterocycle **72** has two C–N bonds to the same carbon atom. If we disconnect both, we get cyclohexanone and a very unstable looking imine **73**. We know how to make imines: combine a carbonyl group with an amine so disconnecting both imines we end up with the diketone **74** and two molecules of ammonia.

Supposing you had not noticed the 1,1-diX relationship but had spotted the imines. Disconnection **72a** takes us directly to the diketone **74** and a very unstable diamine **75**. Now you can't avoid the 1,1-diX disconnections **75** and we get the same starting materials whichever analysis we follow.

But what about the synthesis? When we are making stable 5- or 6-membered rings, syntheses are often very forgiving as you will discover in chapters 29 and 39. All you need to do is to mix together the two ketones with ammonium acetate, to provide both a source of ammonia and an acid catalyst, and **TM72** is formed in good yield.[7]

If the heteroatoms are different and one of them is oxygen, it makes more sense to disconnect the other so that the oxygen of the carbonyl group remains. The phosphonate **76** is an example. The nucleophilic synthon is **77** and this can be made by deprotonation of **78**.

As this chemistry may be unfamiliar, you may like to know that **78**; R=Et, known as diethyl phosphite, is available and forms the anion **77**, better drawn as **79**, with bases and adds **80** to aldehydes to give the anion **81** of the TM.

A real life example is made from **82** and **78**; R=Et with the weak base Et$_3$N. This produces some of the anion **79**; R=Et and the ammonium salt protonates the anion of the TM to give **83** in excellent yield.

Two-Group C–X Disconnections as a Preliminary to a Full Analysis

A brief inspection of the polycyclic cage structure of the natural product sarracenin **84** makes it appear a formidable target for synthesis. As we move forward into the book, it will become more and more important to identify any continuous pieces of carbon skeleton and an essential preliminary for that is to disconnect any structural C-X bonds, preferably using two-group disconnections. That strategy works spectacularly well here. Sarracenin **84** has several C–O bonds in its skeleton. One obvious 1,1-diO relationship is marked with a black blob in **84** showing where an acetal indicates a hidden carbonyl group. The black blob on **84** is the aldehyde in **85**. Disconnecting the acetal gives **85** with two fewer rings.

Further into the skeleton is another hidden carbonyl group **85** masked as a hemiacetal rather than an acetal. Disconnection there shows up an enol **86** and conversion of the enol into the aldehyde gives the simplest structure we have yet seen **87** without any rings at all. Indeed if we redraw that structure in a more conventional way **88**, we can see that it is one continuous piece of carbon skeleton. One published synthesis[8] reconnects the two aldehydes on the right to give one alkene and the aldehyde and alcohol on the left to give another **89**. This compound looks a great deal simpler than sarracenin but in fact has exactly the same number of carbon atoms. We shall meet the reconnection strategy in chapter 27.

References

1. T. Nakajima, S. Masuda, S. Nakashima, T. Kondo, Y. Nakamoto and S. Suga, *Bull. Chem. Soc. Jpn.*, 1979, **52**, 2377.
2. M. Freifelder, *J. Am. Chem. Soc.*, 1960, **82**, 2386.
3. J. L. J. Kachinsky and R. G. Salomon, *Tetrahedron Lett.*, 1977, 3235.
4. J. B. Hendrickson and C. Kandall, *Tetrahedron Lett.*, 1970, 343; W. L. Judefind and E. E. Reid, *J. Am. Chem. Soc.*, 1920, **42**, 1043.
5. K. Richardson, *Contemporary Organic Synthesis*, 1996, **3**, 125.
6. *Sandwich Drug Discoveries*, Pfizer technical booklet, 1999.
7. E. J. Corey, R. Imwinkelried, S. Pikul and Y. B. Xiang, *J. Am. Chem. Soc.*, 1989, **111**, 5493; E. J. Corey, D.-H. Lee and S. Sarshar, *Tetrahedron: Asymmetry*, 1995, **6**, 3; S. Pikul and E. J. Corey, *Org. Synth.*, 1993, **71**, 22.
8. M.-Y. Chang, C.-P. Chang, W.-K. Yin and N.-C. Chang, *J. Org. Chem.*, 1997, **62**, 641.

7 Strategy III: Reversal of Polarity, Cyclisations, Summary of Strategy

This chapter considers in more depth two strategic points that emerged from the discussion of C–X disconnections in chapters 4–6.

Reversal of Polarity
Synthesis of Epoxides and α-Halo-Carbonyl Compounds

In chapter 6 we needed three types of synthon depending on the di-X relationship in the target molecule. For the 1,3-diX relationship we used just one synthon **2**, for the 1,2-diX we used related synthons **5** and **8**, and for the 1,1-diX two more **11** and **14**. The synthons for the 1,3-diX and 1,1-diX relationships could be turned into reagents **3**, **12** and **15** simply by using the natural electrophilic behaviour of the carbonyl group. The synthons **5** and **8** for the 1,2-diX relationship could not be turned into reagents so easily: reagent **6** does not resemble synthon **5** while synthon **8** looks very unstable and such intermediates cannot be made.

We solved those problems by using an epoxide **6** for synthon **5** and an α-haloketone **9** for **8**: two apparently different devices that actually rely on the same principle—one that is the subject of this chapter. It is easy to see with the synthon **8**: if we simply reverse the polarity to the anion

Organic Synthesis: The Disconnection Approach. Second Edition   Stuart Warren and Paul Wyatt
© 2008 John Wiley & Sons, Ltd

16 we discover a synthon that again uses the natural reactivity of the carbonyl group as an enol **17** (or enolate) in equilibrium with the ketone **18** by tautomerisation.[1] Treatment of the ketone **18** with bromine in acidic solution gives the α-haloketone **9** with an electrophilic carbon atom in the right place.

The epoxide **6** is naturally electrophilic, but where does the epoxide come from? By far the most important method of epoxide synthesis is the treatment of alkenes **19** with peroxy acids RCO_3H **21**. Alkenes are naturally nucleophilic:[2] they react with bromine to give dibromides 20 and with electrophilic peroxyacids **21** to give epoxides. Again, these reactions convert nucleophilic alkenes into electrophilic derivatives. A very popular reagent for epoxidation is *m*CPBA (*meta*-chloro-perbenzoic acid) **21**; R = 3-chlorophenyl but many other compounds are used.

The Halogenation of Ketones

The halogenation of ketones must be carried out in acid solution to avoid polyhalogenation.[1] So the synthesis of reagent **22**, used to make derivatives of carboxylic acids in chapter 6, is simple providing that we notice the directing effects of the two groups on the benzene ring in **23** and disconnect with Friedel-Crafts in mind.

The synthesis is very straightforward: no bromination occurs on the ring as would be expected in the absence of a Lewis acid. Enols react with bromine without the need of any catalysis.[3]

This bromination was unambiguous as the ketone could enolise on one side only. In general the reaction is suitable only for ketones that are symmetrical[4] (e.g. **25**), blocked on one side[5] (e.g. **23** or **27**) or which enolise regio-selectively[6] (e.g. **29**).

Halogenation of Acids

There is no ambiguity in the halogenation of acids as they can of course enolise on one side only. Reliable methods are bromination with PCl_3 and bromine or red phosphorus and bromine. The acid is converted into the acyl chloride with PCl_3 or the acid bromide by PBr_3, formed in the reaction mixture from red phosphorus and bromine. Bromohexanoic acid **34** can be made in good yield if the reaction mixture is worked up with water.[7]

If the reaction is quenched with an alcohol, only the acyl halide reacts and this is a simple way to make α-bromoesters **38**. The alternative product **39** is not formed. Water and alcohols are poor nucleophiles in the S_N2 reaction but better with carbonyl groups.

Many α-chloroacids are available commercially (chloro-acetic, propanoic, etc.) and chloroacetyl chloride **41** is made on a very large scale industrially.[8] The α-chloro amide **40**, needed to make some analeptic tetrazoles, is best disconnected as an amide as **41** is cheap.[9]

It is better to acylate aniline before nitration to prevent oxidation or over-nitration and reduce the proportion of *ortho*-nitration. The yield of **45** is after separation from the *ortho* isomer.[10] Notice that in the last step nitrogen, like oxygen, prefers to attack the acyl rather than the alkyl chloride.

Cyclisation Reactions

Generally intramolecular reactions are easier than intermolecular reactions: entropy being a major factor. If you want to make an acetal from a ketone (chapter 6) it is better to use a diol **47** rather than, say methanol. The equilibrium is in favour of the cyclic acetal **48** but not in favour of the methyl acetal **46**. Two molecules—one of each—go into **48** but three—two alcohols and one ketone—go into **46**. Entropy is a thermodynamic factor.

But the rates of cyclisations to form 3-, 5-, 6- and 7-membered rings are greater than the rates of corresponding bimolecular reactions. This is kinetics but the smaller loss of entropy (fewer degrees of freedom lost in the cyclisation) is also a factor. We should not expect a good yield in an acid-catalysed ether formation from two alcohols. If the reaction worked at all, we should get dimers of each alcohol as well as the mixed ether **51**.

But if the reaction were a cyclisation of the diol then things would be very different. The rate of the cyclisation will be much greater so even this unpromising reaction should go well. And no regioselectivity problems would arise. If the side chains on nitrogen were different **52** we should still get the same product **53** regardless of which OH group were protonated and which acted as the nucleophile. The parent compound **54** is morpholine and this unit is present in many drugs such as the analgesic phenadoxone[11] **55**.

The necessary diol for such compounds comes, by two 1,2-diX disconnections **56**, from the amine RNH_2 and two molecules of ethylene oxide. Now we want the epoxide to react twice so an excess is used and the diol **56** cyclised in acid.[12]

Choosing cyclisation reactions can make possible syntheses we should certainly reject if a bimolecular reaction were required. The ether disconnection **58b** gives a perfectly reasonable diol **59** that would certainly cyclise as we wish. But making **59** would involve creating an *ortho* relationship with the unwanted *para* relationship probably preferred.

We should not normally consider the Friedel-Crafts alternative **58a** as the intermediate **61** would be unstable. But what that means is that **61** will cyclise rapidly to **58**. Indeed it is difficult to isolate[13] **61** as it gives **58** even at 35 °C.

A dramatic example occurs in the last stage of the production of sildenafil **63** the Pfizer treatment of male erectile dysfunction better known as Viagra™. The cyclisation of **62** must involve the attack of the nitrogen atom of one amide on the carbonyl atom of the other (arrows show first stage). This is an exceptionally difficult reaction: amides are very poor nucleophiles and very poor electrophiles. Yet this reaction goes in over 90% yield.[14] It does so because it is intramolecular.

The diamide **62** is heated under reflux for several hours in *t*-butanol with the base *t*-BuOK as catalyst so it may be that the anion of the nucleophilic amine is involved. Afterwards, dilution with water and neutralisation to pH 7.5 with HCl gives pure **63**.

Summary of Strategy

In chapter 1 we gave the bare bones of synthetic strategy. We can now add life to those bones by adding the main points from chapters 2–7. There will be fuller outlines as the book progresses.

Analysis:

1. Recognise the functional groups in the target molecule.
2. Disconnect with known reliable reactions in mind, using FGI as needed to give the right FG. Disconnect:
   (a) Bonds joining an aromatic ring to the rest of the molecule, whether Ar–X or Ar–C (chapters 2 and 3);
   (b) Any C–X bond (chapter 4) especially:
       (i) Bonds next to carbonyl groups RCO–X (chapter 4);
       (ii) Using two-group disconnections (chapter 6);
       (iii) Bonds in saturated rings as cyclisations are so good (chapter 7).
3. Repeat as needed to reach available starting materials.

Synthesis:

1. Write out the plan in the forward direction adding reagents and conditions.
2. Check that a rational order of events has been chosen (chapter 3).
3. Check that chemoselectivity is favourable (chapter 5). Use protection if necessary (chapter 9).
4. Modify the plan from points 2 and 3 (and later from unexpected failures or successes in the laboratory).

Example: Salbutamol

The anti-asthma drug salbutamol **64**, better known as GSK's Ventolin™, is closely related to adrenaline **65**. The extra carbon atom, marked with a black blob in **64**, prevents dangerous side effects on the heart and the *t*-butyl group makes the drug longer lasting.

64; salbutamol   **65; adrenaline**

Salbutamol has three hydroxyl groups and an amine but the only two-group C–X disconnection is of the C–N bond **64a** revealing the epoxide **66** as a starting material. This approach is successful but it involves chemistry we encounter in chapter 30 so we shall discuss it there.

64a   **66**

An alternative is FGI back to the ketone **67** and hence the α-bromoketone **68** that can be made from the ketone **69** itself by methods discussed earlier in this chapter. The ketone **69** is clearly made by some sort of Friedel-Crafts acylation, but how are we to make the diol **70**? In chapter 3 we said that a good strategy to make *ortho*-disubstituted aromatic compounds was to start with an available compound with that relationship already present. Here the obvious candidate is salicylic acid **71**.

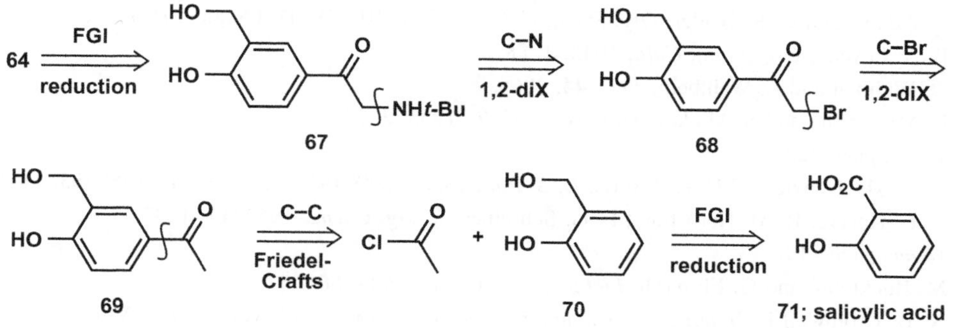

Further consultation of chapter 3 reveals the synthesis of the related ketone **73** by a Friedel-Crafts style reaction on aspirin **72**. As we have two reductions (of the acid and the ketone) it makes sense to do them both at the end. The plan is now:

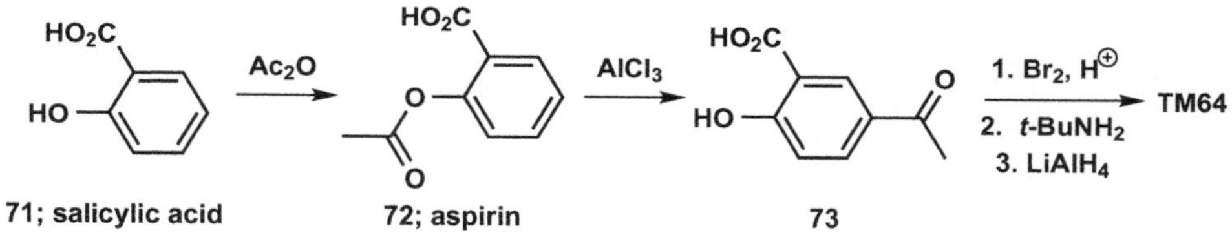

Checking for chemoselectivity problems, we might suspect that the amine could be alkylated twice by the very reactive α-bromoketone **74** so it might be better to protect the nitrogen atom with a benzyl group. This can be removed by catalytic hydrogenation. In the laboratory, it proved better to brominate **73** in neutral rather than acidic solution so the final scheme becomes:

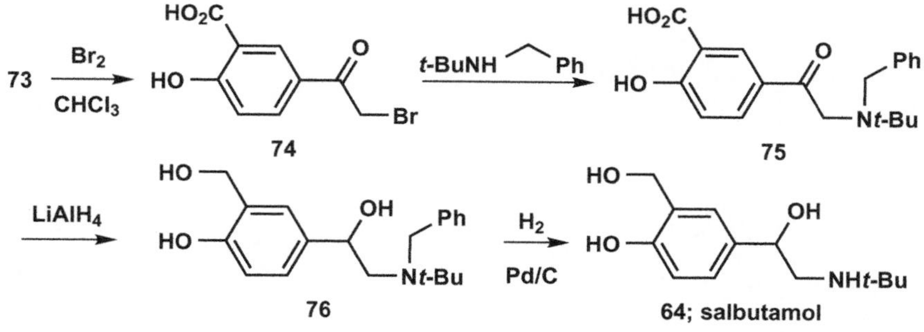

This synthesis is short and high yielding, makes good use of the strategic points used so far in this book and introduces the subjects of the next two chapters: amine synthesis and the use of protecting groups.

References

1. Clayden *Organic Chemistry*, chapter 21.
2. Clayden *Organic Chemistry*, chapter 20.
3. R. Adams and C. R. Noller, *Org. Synth. Coll.*, 1932, **1**, 109; W. D. Langley, *Ibid.*, 127.
4. P. A. Levene, *Org. Synth. Coll.*, 1943, **2**, 88.
5. O. Widman and E. Wahlberg, *Ber.*, **44**, 2065.
6. E. M. Schultz and S. Mickey, *Org. Synth. Coll.*, 1955, **3**, 343.
7. *Vogel*, page 722.
8. G. F. MacKenzie and E. K. Morris, *U. S. Pat.*, 1858, 2,848,491; *Chem. Abstr.*, 1959, **53**, 1151b.
9. E. K. Harvill, R. M. Herbst and E. G. Schreiner, *J. Org. Chem.*, 1952, **17**, 1597.
10. *Vogel*, page 919.
11. M. Bockmühl and G. Ehrhardt, *Liebig's Ann. Chem.*, 1948, **561**, 52.
12. N. H. Cromwell in *Heterocyclic Compounds*, ed. R. C. Elderfield, Vol 6, 1957, Wiley, New York, pp. 502– 517.
13. A. Reiche and E. Schmitz, *Chem. Ber.*, 1956, **89**, 1254.
14. D. J. Dale, P. J. Dunn, C. Golightly, M. L. Hughes, P. C. Levett, A. K. Pearce, P. M. Searle, G. Ward and A. S. Wood, *Org. Process Res. Dev.*, 2000, **4**, 17.

8 Amine Synthesis

Background Needed for this Chapter Reference to Clayden, *Organic Chemistry:* Chapter 14: Nucleophilic Substitution at C=O with Loss of Carbonyl Oxygen.

Amine synthesis needs a separate chapter because the C–X disconnection **1a** used for ethers, sulfides and the like in chapter 4 is not suitable for amines. The problem is that the product of the first alkylation **2** is at least as nucleophilic as the starting material **1** (if not more so because of the electron-donating effect of each alkyl group) and further alkylation occurs giving the tertiary amine **3** or even the quaternary ammonium salt **4**. It is no use adding just one equivalent of MeI as the first formed product **1** will compete with the starting material **2** for MeI.

The simple alkylation of an amine with an alkyl halide can occasionally be used if the product is *less* nucleophilic than the starting material. This may be for electronic reasons: glycine **6** can be made by alkylation of ammonia with **5** as it exists mostly as the zwitterion **7** which no longer has a nucleophilic nitrogen. It may be for steric reasons: the alkylation of the α-bromoketone **8**, mentioned at the end of chapter 7, with the sterically hindered amine **9** gives a good yield of the even more sterically hindered amine **10** and no quaternary salt is formed. If the reaction is a cyclisation (chapter 7) it may also work well.

Organic Synthesis: The Disconnection Approach. Second Edition   Stuart Warren and Paul Wyatt
© 2008 John Wiley & Sons, Ltd

More general solutions come from the replacement of alkylations by reactions with carbonyl compounds. These generally occur once only and in many cases cannot occur twice as the products—amides **12** or imines **15** for example—are much less nucleophilic than the starting amine. The products are reduced to the target amines. The amide route is restricted to amines with a CH_2 group next to nitrogen **13** but the imine route is very general and is known as reductive amination.[1] It is the most important way to make amines and a recent survey showed that the majority of amines made in the pharmaceutical industry are made this way.

A preliminary FGI is needed before we apply the C–N disconnection. Amine **17** could be made from amide **18** or imine **21** and hence from two different primary amines **20** or **22** and two different carbonyl compounds **19** or **23**. These methods are very versatile.

One published synthesis of this amine **17** is by reductive amination.[2] Note that it is not necessary, nor usually desirable, to isolate the rather unstable imine as reduction with $NaB(CN)H_3$ or $NaB(OAc)_3H$ occurs under the conditions of imine formation.[3] Since the imine is in equilibrium with the starting materials, slightly acidic conditions must be used so that the protonated imine is reduced more rapidly than the aldehyde or ketone. These two reducing agents are stable down to about pH 5.

An example that has been made by the amide route is the cyclic amine **24**. Putting the carbonyl group on the side chain **25** allows us to use readily available piperidine **26** as a starting material. The synthesis[4] uses catalytic reduction to give **24** in 92% yield from the amide **25**.

Reductive Amination

This most versatile of amine syntheses can be used to make primary, secondary or tertiary amines providing only that an imine can be formed with an aldehyde or ketone. But tertiary carbon atoms cannot be joined to nitrogen by reductive amination as a tertiary carbon atom cannot have a carbonyl group. The method works by selective reduction of the imine **28** in the presence of the aldehyde **27** or ketone. Catalytic hydrogenation reduces the imine **28** preferentially as the C=N bond of the imine is weaker than the C=O bond of the aldehyde or ketone.

Normal nucleophilic reducing agents like $NaBH_4$ would reduce the more electrophilic aldehyde **27** in preference to the imine **28**. They must be used in slightly acidic solution (pH 5–6) so that the more electrophilic imine salt **29** is reduced. But reducing agents like $NaBH_4$ are unstable in acidic solution, decomposing to hydrogen gas. That is why modified borohydrides [$NaB(CN)H_3$ or $NaB(OAc)_3H$] are used. The electron-withdrawing CN or OAc groups reduce the nucleophilicity of the hydride(s) attached to boron, making it more selective towards the imine salt **29** and stabilising it in acid.

If the imine is stable enough to be isolated,[5] as with diaryl imines **32** or crowded aliphatic amines such as **35**, then $NaBH_4$ can be used for the reduction as there is no competition with unreacted aldehyde.

Primary Amines by Reductive Amination

The amine needed would be ammonia but unsubstituted imines **36** are very unstable. Ammonium acetate is usually used as the source of ammonia and to get the right pH for reductive amination with $NaB(CN)H_3$ or $NaB(OAc)_3H$. Either aldehydes **37**; $R^2 = H$ or ketones **37** can be used.

Secondary Amines by Reductive Amination

Examples **17**, **30** and **33** show how this works with aldehydes. Ketones give amines such as **40** and both can be discovered just by using the disconnections **28a** and **41**. If one of the two carbon atoms joined to nitrogen is tertiary, that must be R^2 in **30** or R^3 in **40** as a tertiary centre cannot be set up by reduction.

Tertiary Amines by Reductive Amination

It may appear at first sight that tertiary amines cannot be made by reductive amination as an imine cannot be made. If a secondary amine such as piperidine **42** reacts with an aldehyde, the product is an enamine **44** not an imine. But reflect: the enamine **44** is formed by deprotonation of the imine salt **45** and that is the species we need for reaction with NaB(CN)H$_3$ or NaB(OAc)$_3$H to give the tertiary amine **46**. So there is no problem.

The disconnections are straightforward: just draw the iminium salt **48** or **50** after FGI on the tertiary amine **47** or **49** and disconnect the C=N bond in the usual way. You will often have three choices as to which iminium salt you draw. Only if one of the substituents on nitrogen is tertiary is that option not available. We explore that problem soon.

Other Ways to Make Amines

Primary Amines by Alkylation with Alkyl Halides

There is one method of direct alkylation of a nitrogen nucleophile. Preliminary FGI (with reduction in mind again) to an alkyl azide **52** allows C–N disconnection to the alkyl halide and azide ion **54**. This interesting species is linear and can slip into crowded molecules like a tiny dart. But there is a drawback: all azides are toxic and POTENTIALLY EXPLOSIVE.

A salt such as sodium azide is used and the reduction can be carried out catalytically, with NaBH$_4$ or with Ph$_3$P in protic solution. Simple amines such as octylamine **57** can be made this way.[6] The azide **56** is not isolated but the whole reaction sequence carried out in the same aqueous solution to reduce the danger of an explosion.

A more deep-seated disconnection comes from a different FGI (using reduction yet again) with the idea that cyanide ion **61** should be the nucleophile. This makes a C–C bond rather than a C–N bond but does at least disconnect two atoms. Cyanide is an interesting structure: it has to be linear and it has a lone pair on nitrogen and a negative charge on carbon making it one of the rare genuine carbanions. There is again a drawback: cyanides are very toxic.

This method is particularly useful if the S_N2 reaction with cyanide is favourable as with benzyl bromide **62**. The reduction can be carried out with a variety of reagents: here hydrogenation over Raney nickel gives a good result.[7]

Joining Tertiary Carbon to Nitrogen

One way to do this uses aliphatic nitro compounds and is discussed in chapter 22. One direct method is the Ritter reaction[8] successful only for tertiary alkyl groups as it involves an S_N1 reaction. The nitrogen nucleophile is a nitrile—a notoriously weak nucleophile that needs a carbocation for reaction. If *t*-butanol and acetonitrile are mixed in acidic solution, the tertiary cation is attacked by the nitrile **66** and the amide **69** is formed. Hydrolysis of the amide gives *t*-BuNH$_2$ or reduction of the amide gives the secondary amine **70**. The nitrile is chosen according to the other alkyl group needed.

The Synthesis of Monomorine I

We end with an example that includes methods from this chapter as well as some revision and a reminder of stereochemistry. Monomorine I **71** is the trail pheromone of Pharaoh's ant (*Monomorium pharaonis*). These ants are pests in hospitals as they spread infections and they follow a trail of monomorine as they go about their evil work. Synthetic monomorine might be used to lure the ants to their doom. It is a bicyclic amine and disconnection at all the C–N bonds with reductive amination in mind reveals a linear triketone, drawn more clearly as **72a**.

71; monomorine   **72**   **72a**

The chemists decided[9] that reacting **72** with ammonia was asking a bit much so they selected the nitro compound **73** as their starting material. The idea is that the nitro group will provide the central nitrogen atom after reduction. As we shall see in chapters 22 and 24 nitro groups stabilise carbanions well and conjugate addition of such anions works well. Hence the disconnection of **73**. Nitropentane **75** might be made by alkylation of the anion of nitropropane **75b** or by the method chosen, an S_N2 reaction of nitrite anion on bromopentane **75a**.

73   **74**   **75**

Now for the synthesis. The nitro compound **75** was made from bromopentane and a nitrite in DMSO, a good solvent for S_N2 reactions, and added to the enone **76**, an acetal derived from the diketone **74** with the strong base tetramethyl guanidine **77** as catalyst to give the partly protected form **78** of **73**. Now all is ready for the various reductions.

76   **78; 64% yield**

Catalytic reduction of the nitro group gives the amine **79** that cyclises instantly (chapter 7) to the imine **80** reduced in its turn to the cyclic amine **81**. When the virtually planar five-membered ring of the imine settles on the surface of the Pd/charcoal catalyst it can choose between the side of the ring with a hydrogen atom or the side with the butyl group. It chooses the less hindered side and so the second hydrogen atom is *cis* to the first and the stereochemistry is correct (compare **81** with **71**).

79   **80**   **81**

Now the acetal is hydrolysed to reveal the ketone **82** which again cyclises spontaneously to the enamine **83** forming a stable six-membered ring. This cyclic enamine can be isolated and treated with $NaB(CN)H_3$ in slightly acidic solution. The enamine is thus in equilibrium with the iminium salt (compare **44** and **45**) and reduction again occurs on the less hindered side of the molecule, i.e. *cis* to the other two hydrogen atoms.

This elegant synthesis uses some of the methods of amine synthesis from this chapter and looks forward to the next chapter on protecting groups as well as later discussion of nitro group chemistry.

References

1. E. W. Baxter and A. B. Reitz, *Org. React.*, 2002, **59**, 1.
2. K. A. Schellenberg, *J. Org. Chem.*, 1963, **28**, 3259.
3. A. F. Abdel-Magid and S. J. Mehrman, *Org. Process Res. Dev.*, 2006, **10**, 971.
4. B. Wojcik and H. Adkins, *J. Am. Chem. Soc.*, 1934, **56**, 2419.
5. *Vogel*, page 783.
6. *Vogel*, page 772.
7. R. F. Nystrom, *J. Am. Chem. Soc.*, 1955, **77**, 2544; Vogel, page 773.
8. L. I. Krimen and D. J. Cota, *Org. React.*, 1969, **17**, 213.
9. R. V. Stevens and A. W. M. Lee, *J. Chem. Soc., Chem. Commun.*, 1982, 102; R. V. Stevens, *Acc. Chem. Res.*, 1984, **17**, 289.

9 Strategy IV: Protecting Groups

Background Needed for this Chapter Reference to Clayden, *Organic Chemistry:*
Chapter 24: Chemoselectivity: Selective Reactions and Protection.

Protecting groups have been mentioned occasionally in previous chapters: in this chapter the ideas behind their use are systematically presented and a collection of protecting groups suitable for a range of functional groups is tabulated. Protection allows us to overcome simple problems of chemoselectivity. It is easy to reduce the keto-ester **1** to the alcohol **2** with a nucleophilic reagent such as NaBH$_4$ that attacks only the more electrophilic ketone.

Making alcohol **3** by reducing the *less* electrophilic ester is not so easy but protection of the ketone as an acetal **4**—a functional group that does not react with nucleophiles—allows reduction of the ester with the more nucleophilic LiAlH$_4$.

Another important function of protecting groups is to prevent a reagent from attacking itself. In the last chapter, when we discussed the synthesis of the bicyclic amine monomorine, we used the protected enone **12** but did not say how it was made. The chloroketone **9** is first made[1] in 89–93% yield from the ketolactone **6** simply by reaction with HCl. Chloride displaces the protonated ester group **7** and the product **8** decarboxylates under the conditions of the reaction.

Organic Synthesis: The Disconnection Approach. Second Edition   Stuart Warren and Paul Wyatt
© 2008 John Wiley & Sons, Ltd

Any attempt to make a Grignard reagent from **9** is doomed because the nucleophilic Grignard would immediately attack the ketone. We need to protect the ketone with an easily added group that is not attacked by Grignard reagents and the acetal **10** is the answer. Addition of the Grignard from **10** to acrolein (CH$_2$=CHCHO) gives the allylic alcohol **11** which is oxidised to the enone **12** with manganese dioxide.[2] If you look back to chapter 8 you will see that the acetal was retained until it was very easily removed almost at the end of the synthesis.

Qualities Needed in a Protecting Group

1. It must be easy to put in.
2. It must be resistant to reagents that would attack the unprotected functional group.
3. It must be easily removed.

The last point may not be so obvious but it is the most difficult to achieve. Many syntheses fall down right at the end because a protecting group cannot be removed without destroying the molecule. The next section looks at ways to make removal of protecting groups easier.

Ethers and Amides as Protecting Groups

Protection of alcohols and amines might look simple. Methyl ethers and simple amides are easy to make and are very resistant to a wide variety of reagents. So there is no problem in carrying out the required reaction elsewhere in the molecule i.e. turning R^1 in **13** and **17** into R^2 in **16** and **20**. But sadly they are almost useless as protecting groups because such violent conditions are needed to remove them: cleavage of methyl ethers requires good nucleophiles under acidic conditions and the hydrolysis of amides needs refluxing 10% NaOH or concentrated HCl in a sealed tube at 100 °C overnight.

These protecting groups are used when the molecule is robust enough to take the deprotection conditions. If aniline **21** is brominated the 2,4,6-tribromo derivative **22** is formed. The yield is quantitative but we are more likely to want mono-bromination. Protection is needed against over-reaction. The amide **23** is easily made, bromination goes only in the *para* position (the *N*-acetyl group is larger than NH$_2$) and the hydrolysis does not destroy the benzene ring.[3]

The Achilles Heel Strategy

A way round these difficulties is to use an ether or an amide that has a built-in weakness so that the over-vigorous conditions are not needed. This 'Achilles heel' for an ether is commonly the THP group that makes the ether into an acetal. Dihydropyran, DHP **26**, is protonated on carbon **27** to give the cation **28** that captures the alcohol to give the mixed acetal **29**, usually referred to as 'the THP derivative'. After the reaction the hydrolysis needs only the weak aqueous acid used for acetals. The secret is that the weak acetal bond (**b** in **30**) is cleaved[4] rather than the strong ether bond (**a** in **30**).

Another way to make an ether easier to remove is to make it benzylic **31** as σ-conjugation of the C–O bond with the benzene ring weakens it enough for it to be cleaved by catalytic hydrogenation using various transition metals.[5]

This is also the key to the weakening of amides as protecting groups for amines. The amine is acylated with benzyl chloroformate **33** (as described in chapter 5) to give urethane **34**. This is still an amide on the left but a benzyl ester on the right. Then the reaction is carried out. Hydrogenation cleaves the weak benzyl-O bond to give unstable carbamic acids **36** that decarboxylate **37** spontaneously to give the altered amine R^2NH_2. Though the C–N bond is cleaved, no nucleophilic attack on the carbonyl group is needed. This protecting group is so popular it has its own abbreviation Cbz (Carbobenzyloxy-) or even just Z.

Both benzyl and Cbz groups are used in a synthesis of aspartame **38**, the dipeptide that is 150 times sweeter than sugar and used in many soft drinks under the name Nutrasweet™. Only one disconnection is reasonable: the amide bond in the middle of the molecule suggesting derivatives of available aspartic acid **39** and phenylalanine **40** as starting materials.

Since we need the methyl ester of phenylalanine, no further protection of that starting material is needed but the amino and one carboxylic acid group of aspartic acid need to be protected and the remaining acid activated. The Cbz group is perfect for the amino group and both acids in **41** can be esterified with benzyl alcohol.

Now one ester must be cleaved and not the other. This looks difficult and cannot easily be done by hydrogenolysis but peptide chemists knew that the right one was hydrolysed in base to give the required intermediate **43**. Evidently the amide makes that ester more electrophilic but this was discovered by experience. Now the free acid must be activated towards nucleophilic attack: the trichlorophenyl ester **44** is ideal.[6]

The coupling requires only a weak base and the benzyl esters are removed by hydrogenation. The benzyl esters are there for protection but the trichlorophenyl ester is there for activation, making that ester more electrophilic than the benzyl ester in **44** or the methyl ester in **40**.

A close relative to Cbz is Boc (*t*-butyloxycarbonyl) that uses a different method to make esters easy to hydrolyse. It is added to an amine or an alcohol by the chloroformate **46** and, after the reaction, 'hydrolysed' with acid—no water being needed. The ester is protonated and the *t*-butyl cation drops out in an S_N1 reaction **49** to give the same intermediate **36** as in the removal of the Cbz group.

Protection of Alcohols

We have already mentioned the THP group but by far the most popular protecting groups for alcohols are the various silyl groups. You will already be familiar with the Me_3Si- or TMS (TriMethylSilyl) group but this is little used for protections as it falls off so easily, often just during chromatography. More hindered is the triethylsilyl (TES) or tri-*iso*-propylsilyl (TIPS) groups and *t*-butyldimethyl silyl (TBDMS or, misleadingly TBS) and, most hindered of them all, *t*-butyldiphenyl silyl (*t*-BuPh$_2$Si-). They are usually put on with the silyl chloride and a weak base, often imidazole in DMF, and they can be removed with oxygen nucleophiles, often in acid solution, and especially fluoride ion, often as TBAF (TetraButylAmmonium Fluoride Bu$_4$NF). This is particularly useful as fluoride is virtually unreactive towards most carbon atoms.

In Martin and Mulzer's synthesis[7] of epothilone B the starting material **50** already has a *p*-methoxybenzyl group on one alcohol. Protection of the other with a TBDMS group is *orthogonal* meaning that each group is removed under conditions that do not affect the other. Addition of isopropenyl Grignard to the aldehyde **52** creates a third alcohol **53** and now the TBAF group is removed so that an acetal **55** can be formed from the diol **54**. The yields are all good and eventually the PMB group will be removed by oxidation with a quinone.

An intermediate in the synthesis of laulimalide by Davidson[8] illustrates the differential protection of alcohols. The starting materials **56** and **57** already have an alcohol protected as a TBDMS silyl ether and a diol protected as an acetal. The alcohol in **58** is protected as a *p*-methoxybenzyl ether and the acetal 'hydrolysed' by acetal exchange to give the free diol **60**. Selective protection of the primary alcohol by a bulky acyl group (pivaloyl, *t*-BuCO-) **61** allows silylation of the secondary alcohol with a TIPS group **62**. Finally the pivaloyl group is selectively removed by DIBAL reduction to release just one free alcohol **63**.

56   **57**   **58; 82% yield**

1. (Me₃Si)₂NK

2. (p-MeO-benzyl)Br

59

HS~SH, Et₂O–BF₃

60; 71% yield from 58

t-BuCOCl, Et₃N, CH₂Cl₂

61; 93% yield

i-Pr₃SiCl, base

62; 95% yield

DIBAL, CH₂Cl₂

63; 98% yield

Later on, all the protecting groups will be removed: the silyl groups with fluoride and the *p*-methoxybenzyl ether by oxidation with Ce(IV). In Ley's recently completed synthesis[9] of azadirachtin **64** after 22 years of hard labour the key intermediate was **65**. You will notice benzyl ethers, acetals and a silyl ether. This is a more modern, one might almost say minimalist, use of protection. In an ideal world no protecting groups would be necessary but in a real synthesis they will almost certainly be required as we shall see in the rest of the book. But our aim should be to keep them to a minimum.

64; azadirachtin   **65; R = *t*-BuMe₂Si**

The Literature on Protecting Groups

'Protecting groups' is a very large subject. There are hundreds of different protecting groups using scores of different ideas for every important functional group. It is particularly important that you refer to the literature before you choose which protecting group to use in a synthesis. It is a reasonable assumption that this rather dull subject would spawn some rather dull catalogue-like textbooks, and it has, but fortunately there is a glorious exception. Phil Kocienski's textbook[10]

Protecting Groups is comprehensive and entertaining. If you doubt this, have a look at page 644 (yes, it's also a long book). The extent of the subject is revealed by his chapter on protecting groups for alcohols with hundreds of protecting groups in the 176 pages and 686 references. This chapter is particularly good at selective protection and deprotection of, say, primary, secondary and tertiary alcohols. One example is the selective deprotection of either a phenol or an alcohol from the *bis*-silyl ether 67 using different reagents.[11]

Protecting Group Summary

We cannot compete with large textbooks but here is a brief selection of simple protecting groups.

TABLE 9.1   Simple Protecting Groups (PGs) for some functional groups

Protecting Group	To Add	To Remove	PG resists	PG reacts with
Protecting Aldehydes RCHO and Ketones R$_2$CO				
Acetal (Ketal)	ROH or diol, H$^+$	H$^+$, H$_2$O	nucleophiles, bases, reducing agents	electrophiles, oxidising agents
Protecting Carboxylic Acids RCO$_2$H				
Ester: RCO$_2$Me	CH$_2$N$_2$	NaOH, H$_2$O		
Ester: RCO$_2$Et	EtOH, H$^+$	NaOH, H$_2$O	bases elec-	strong bases
Ester: RCO$_2$Bn	BnOH, H$^+$	H$_2$, cat or HBr	trophiles	nucleophiles
Ester: RCO$_2$t-Bu	t-BuOH, H$^+$	H$^+$		
Anion: RCO$_2$$^-$	base	acid	nucleophiles	electrophiles
Protecting Alcohols ROH				
Ether: ROBn	PhCH$_2$Br, base	H$_2$, cat or HBr	see text	nucleophiles
Silyl ether	R$_3$SiCl, base	F$^-$ or H$^+$, H$_2$O	see text	nucleophiles
Acetal: THP	DHP, H$^+$	H$^+$, H$_2$O	bases	acids
Ester: ROCOR'	R'COCl, pyr	NH$_3$, MeOH	electrophiles	nucleophiles
Protecting Phenols ArOH				
Ether: ArOMe	Me$_2$CO$_3$, K$_2$CO$_3$	HI, HBr or BBr$_3$	bases	electrophiles
ArOCH$_2$OMe	MeOCH$_2$Cl/ base	HOAc, H$_2$O	bases	electrophiles
Protecting Amines RNH$_2$				
Amides RNHCOR'	RCOCl	NaOH or HCl in water	electrophiles	bases and nucleophiles
Urethanes RNHCO$_2$R'	ROCOCl	see text	electrophiles	bases and nucleophiles
Protecting Thiols RSH				
Thioester RSAc	AcCl, base	NaOH, H$_2$O	electrophiles	oxidation

References

1. G. W. Cannon, R. C. Ellis and J. R. Leal, *Org. Synth.*, 1951, **31**, 74.
2. R. V. Stevens and A. W. M. Lee, *J. Chem. Soc., Chem. Commun.*, 1982, 102;   R. V. Stevens, *Acc. Chem. Res.*, 1984, **17**, 289.
3. *Vogel*, pages 909, 918.
4. *Vogel* page 552.
5. W. H. Hartung and R. Simonoff, *Org. React.*, 1953, **7**, 263.
6. R. H. Mazur, J. M. Schlatter and A. H. Golgkamp, *J. Am. Chem. Soc.*, 1969, **91**, 2684.
7. H. J. Martin, P. Pojarliev, H. Kählig and J. Mulzer, *Chem. Eur. J.*, 2001, **7**, 2261.
8. A. Sivaramakrishnan, G. T. Nadolski, I. A. McAlexander and B. S. Davidson, *Tetrahedron Lett.*, 2002, **43**, 213.
9. G. E. Veitch, E. Beckmann, B. J. Burke, A. Boyer, S. L. Maslen and S. V. Ley, *Angew. Chem. Int. Ed.*, 2007, **46**, 7629.
10. P. J. Kocienski, *Protecting Groups*, 3rd Edition, Thieme, Stuttgart, 2004.
11. E. W. Collington, H. Finch and I. J. Smith, *Tetrahedron Lett.*, 1985, **26**, 681.

10 One Group C–C Disconnections I: Alcohols

Background Needed for this Chapter Reference to Clayden, *Organic Chemistry:* Chapter 9: Using Organometallic Reagents to Make C–C Bonds.

We now leave disconnections of bonds between carbon and other atoms (C–X disconnections) and turn to the more challenging C–C disconnections. These are more challenging because organic compounds contain many C–C bonds and it is not clear at first which ones should be disconnected. There is some very good news: the synthons that we met in chapter 6 for two-group C–X disconnections are the ones we shall use for one-group C–C disconnections. We start with an introduction to the three main types. In each case we shall replace one of the heteroatoms by a carbon unit 'R'.

For compounds with two heteroatoms joined to the same carbon, we used a 1,1-diX disconnection **1** removing one heteroatom to reveal a carbonyl compound, here an aldehyde, and a heteroatom nucleophile **2**. Replacing the heteroatom by R^2, we disconnect in the same way to reveal the same aldehyde and some nucleophilic carbon reagent **4**, probably R^2Li or R^2MgBr.

1,1-diX Disconnections: **The Corresponding C–C Disconnection:**

For compounds with a 1,2-relationship **5** we used an epoxide **6** at the alcohol oxidation level in combination with a heteroatom nucleophile. Disconnecting the corresponding C–C bond **7**, we use the same epoxide and a carbon nucleophile such as RLi or RMgBr.

1,2-diX Disconnections: **The Corresponding C–C Disconnection:**

The same 1,2-diX relationship at the carbonyl level was disconnected **8** to give carbon electrophile **9**, probably an α-bromoketone, and a heteroatom nucleophile. Now we come to some more good news. We generally preferred nucleophilic heteroatoms but we can use nucleophilic or electrophilic carbon atoms whichever is better. Here we should much rather use the nucleophilic carbon synthon **11** as it is an enolate.

Organic Synthesis: The Disconnection Approach. Second Edition   Stuart Warren and Paul Wyatt
© 2008 John Wiley & Sons, Ltd

The 1,3-diX relationship **12** was quickly recognised as conjugate addition to the enone **13** in chapter 6. The corresponding C–C disconnection **14** uses the same enone **13** but the nucleophilic carbon species should be a copper derivative: RCu, R$_2$CuLi or RMgBr with Cu(I)Br.

Reagents for Nucleophilic Carbon

The simplest unfunctionalised carbon nucleophiles (**15** and **17**) are made from alkyl halides with various metals such as Li(0) or Mg(0) or by exchange with available organometallic reagents such as butyl-lithium (BuLi) in anhydrous coordinating solvents like ether (Et$_2$O) or THF (tetrahydrofuran **16**). Enolates **11** are very important and will be discussed at length in later chapters.

'1,1 C–C' Disconnections: The Synthesis of Alcohols

Disconnection **3** shows that any alcohol may be disconnected at a bond next to the OH group. Isomeric alcohols **18** and **20** can both be made from acetone using perhaps a Grignard reagent **19** in the first case and available BuLi in the second.

The synthesis of **18** exemplifies the Grignard method. The reagent is made[1] from the alkyl halide with magnesium metal in dry ether and combined, without isolation, with the electrophile— all steps being carried out under strictly anhydrous conditions.

It may be necessary to disconnect structural C–X bonds before doing the C–C disconnection as with the aminoester wanted for evaluation as an analgesic.[2] Disconnecting the ester reveals the tertiary alcohol **23** and removal of the phenyl group shows a hidden 1,3-diX relationship between ketone and amino groups **24**.

The synthesis is straightforward with available PhLi being used instead of a Grignard. The acylation of the tertiary benzylic alcohol **23** needs mild conditions to avoid dehydration.

In general there is a choice of which C–C bond should be disconnected and available starting materials may give a clue. We do not wish to disconnect the aromatic ring of the heterocyclic alcohol **28** so we can choose between bonds **a** and **b**.

It turns out that both the aldehyde **29** and the easily made bromo-acetal **32** are commercially available and so route **b** was chosen[3] with the protected Grignard reagent **33** as the carbon nucleophile (compare compound **10** in chapter 9).

Aldehydes and Ketones

The simplest route to aldehydes and ketones using the same strategy is oxidation of an alcohol. So the analysis involves FGI back to the alcohol and then a C–C disconnection of one of the bonds next to the OH group. Lythgoe[4] wanted to make a series of ketones **34** with various R groups to demonstrate a new alkyne synthesis. Disconnection of the C–R bond of the alcohol **35** meant that they could all be made from aldehyde **36** which can be made by the same strategy.

The oxidation of **35** presents no problems as over-oxidation cannot occur. But aldehyde **36** could be oxidised to the corresponding carboxylic acid so care was needed. In fact PCC (pyridinium chlorochromate: CrO_3 and HCl dissolved in pyridine) could be used for both.[5]

Direct addition of RMgBr or RLi to esters does not give ketones (see below) but addition to nitriles does[6] (chapter 13).

Oxidising Agents for the Conversion of Alcohols to Aldehydes

The difficulty is over-oxidation. One simple solution is to oxidise all the way to the carboxylic acid and reduce selectively with, say DIBAL (*i*-Bu$_2$AlH). But the reagents in the table give reasonable results and can also be used for the oxidation of secondary alcohols to ketones.[7] Full descriptions are in Fieser[8] or the volume of *Comprehensive Organic Synthesis* devoted to oxidation.[9]

TABLE 10.1   Reagents for oxidising alcohols to aldehydes

Name	Reagents	For RCH$_2$OH to RCHO
–	$Na_2Cr_2O_7$, H^+	distil out RCHO as formed
Jones	CrO_3, H_2SO_4, acetone	distil out RCHO as formed
Collins	CrO_3, pyridine	use in CH_2Cl_2 solution
PCC	CrO_3, pyridine.HCl	no modification needed
PDC	(pyridine.H^+)$_2$ Cr_2O_7	use in CH_2Cl_2 solution
Swern	1. $(COCl)_2$, DMSO, 2. Et_3N	no modification needed

References for table: $Na_2Cr_2O_7$, H^+: *Vogel*, p. 588, Collins,[10] PCC,[11] PDC,[11] Swern.[12]

Carboxylic Acids

The same disconnection **41** can be used for carboxylic acids with CO_2 as the electrophile for a Grignard reagent **40**. Dry ice (solid CO_2) is particularly convenient for these reactions. Switching polarity by FGI to the nitrile **42**, the same disconnection now uses cyanide ion as the nucleophile but the same alkyl halide **39** that was used to make the Grignard reagent. Mechanistic considerations should decide between these alternatives.

If the carboxyl group is attached to a tertiary, or even a secondary carbon atom, the S_N2 reaction with cyanide will not be so good and the carboxylation of a Grignard reagent is probably better. Pivalic acid **44** is available but can be made from *t*-BuCl in good yield.[13] A detailed procedure[14] for the acid **46** describes how the acid is extracted from the ether with aqueous NaOH, separated from the water by neutralisation with HCl, and distilled.

If on the other hand the S$_N$2 reaction with cyanide is favoured, as with allylic **47** or benzylic **50** halides, that method is better.[15] Hydrolysis of the nitrile **48** gives the acid **49** but treatment with an alcohol in acidic solution gives the ester **52** directly.[16]

Acids can also be made by the oxidation of alcohols and acid derivatives are available from the acids via the acid chloride. Since acids can also be reduced to alcohols, there is a great deal of interdependence in all these methods. The synthesis of carbonyl compounds by one-group C–C disconnections is discussed more fully in chapter 13.

'1,2 C–C' Disconnections: The Synthesis of Alcohols

The analogy between this type of C–C disconnection and 1,2-diX disconnections was explained at the start of this chapter with compounds **5, 6** and **7**. The epoxide route works particularly well if the epoxide is mono-substituted as the reaction with nucleophilic carbon should then be regioselective. Alcohol **53** is used in perfumery and can be disconnected **53a** at the next-but-one bond to the alcohol group with the idea of using the epoxide **54** made from the but-1-ene.

A Grignard or organo-lithium reagent would attack at the less hindered end of the epoxide and the Grignard route gives the alcohol **53**. In chapter 12 we shall see that this reaction is stereospecific.

We shall not extend the discussion on the 1,2-style of C–C disconnection as it is treated extensively, particularly at the carbonyl oxidation level, later in the book. We simply offer a table of the large number of derivatives that can be made from the alcohols we have been discussing in this and previous chapters. In all these cases, the first step would be FGI to the alcohol and then a C–C disconnection could be chosen.

TABLE 10.2 Compounds made from alcohols

Reaction Type	Product	Chap	Further Products	Chap
Oxidation	aldehydes ketones acids	10	amines by reductive amination or reduction of amides	8
Esterification	esters	4	amines by reduction of amides	8
Tosylation	ROTs	4	other substitutions (see below)	4
HBr or PBr$_3$	bromides		ethers, sulfides	4
		4	thiols	5
SOCl$_2$	chlorides		nitriles	10

Example of the Synthesis of Alcohols and Related Compounds

The alcohol **55** was needed for the synthesis of a bicyclic amine. Disconnection either side of the alcohol gives the aldehyde **57** and the Grignard **56** as starting materials.[17] But could we not also disconnect the other side as well?

Symmetrical alcohols can in fact be made in one step from Grignard reagents and esters, as the reaction first produces the aldehyde **57** which is more electrophilic than the ester and so reacts again. There is a warning here! Aldehydes cannot be made by acylation of Grignard reagents with esters. But if two reactions are wanted, this is a good method.

The tertiary chloride **58** was needed for a study of the effects of electron-withdrawing groups on the S$_N$1 reaction. FGI to the alcohol **59** suggests a C–C disconnection to a Grignard reagent **60** and acetone.

The nitro group must be introduced at some stage and the other substituent is always large and *ortho, para*-directing so it doesn't seem to matter when. As they wanted to make a series of compounds with electron-withdrawing groups on the benzene ring, they chose to make **62** as a common intermediate and nitrate last.[18]

Darifenacin **63** is Pfizer's treatment for urinary urge incontinence. Disconnection at the C–N bond with some amine synthesis in mind (chapter 8) gives a much smaller heterocycle **64** that can again be disconnected in the middle with the idea of alkylating some enolate such as **65** with the derivative of an alcohol **66**. This is attractive because **66** is available as a single enantiomer cheaply from the amino acid hydroxyproline.[19]

There are two problems. Enolates of primary amides are not very practical as the NH protons are more acidic than the CH protons. The solution is to use the nitrile and hydrolyse it later to the amide. A more serious problem is that the S_N2 reaction we want to use to couple the two together will go with inversion and that will give the biologically inactive enantiomer of darifenacin. The solution is a double inversion. Protection of the amine by tosylation **67** is followed by tosylation of the alcohol with inversion using a Mitsunobu-style reaction. This unusual esterification goes reliably with inversion.[20]

The nitrile **70** gives a stabilised anion with NaH that reacts with the tosylate with inversion as expected. The rather unusual sulfonamide deprotection with HBr in phenol gave the amine **72** that was coupled to the rest of the molecule as an amide. Reduction of the amide to the amine and, finally, hydrolysis of the nitrile to the amide gave darifenacin **63**.

Other One-Group C–C Disconnections

There are many other reactions that make C–C bonds using only one functional group. Among the most important involve alkynes by alkylation **73** (chapter 16), alkenes by the Wittig reaction **74** (chapter 15) and nitro compounds by alkylation **75** (chapter 22). Disconnections of alkenes outside the double bond **76** and especially disconnections of dienes between the double bonds **77** use palladium chemistry and are discussed extensively in *Strategy and Control*.

Carbon–Carbon Disconnections to Avoid

All the disconnections we have mentioned use functional groups to guide us. Nowhere will you find the disconnection of one alkyl group from another **78** without any functionality. It might seem that the reaction of a Grignard reagent **79** with an alkyl halide **80** would make **78**, and so it might. But these species will be in equilibrium with **81** and **82**. So, even if the coupling does work, we would get a mixture of **78** and both dimers. It is very much better to let functional groups guide your disconnections.

$$R^1 \!\!\not\!\!- R^2 \implies R^1\!-\!MgBr \;+\; Br\!-\!R^2 \rightleftharpoons R^1\!-\!Br \;+\; BrMg\!-\!R^2$$
$$\quad\; \textbf{78} \qquad\qquad \textbf{79} \qquad\quad \textbf{80} \qquad\qquad\quad \textbf{81} \qquad\quad \textbf{82}$$

References

1. *Vogel*, pages 537–541.
2. M. G. Mertes, P. Hanna and A. A. Ramsey, *J. Med. Chem.*, 1970, **13**, 125.
3. H. Muratake and M. Natsume, *Heterocycles*, 1989, **29**, 783.
4. B. Lythgoe and I. Waterhouse, *J. Chem. Soc., Perkin Trans. 1*, 1979, 2429.
5. E. J. Corey and J. W. Suggs, *Tetrahedron Lett.*, 1975, 2647.
6. M. S. Kharasch and O. Reinmuth, *Grignard Reactions of Non-Metallic Substances*, Prentice-Hall, New York, 1954, pages 767–845.
7. *Vogel*, pages 607–611.
8. Fieser, *Reagents*.
9. *Comprehensive Organic Synthesis*, eds B. M. Trost and Ian Fleming, Pergamon, Oxford, 1991, volume **7**, Oxidation, ed. S. V. Ley.
10. J. C. Collins, W. W. Hes and F. J. Franck, *Tetrahedron Lett.*, 1968, 3363.
11. E. J. Corey and G. Schmidt, *Tetrahedron Lett.*, 1979, 399.
12. A. J. Mancuso and D. Swern, *Synthesis*, 1981, 165.
13. Ref. 6, pages 913–960.
14. *Vogel*, page 674.
15. K. Friedrich and K. Wallenfels in *The Chemistry of the Cyano Group*, ed. Z, Rappoport, Interscience, London, 1970, pp. 67–110;   F. C. Schaeffer, *Ibid.*, pages 256–262.
16. R. Adams and A. F. Thal, *Org. Synth. Coll.*, 1932, **1**, 107, 270;   J. V. Supniewsky and P. L. Salzberg, *Ibid.*, 46; E. Reitz, *Ibid.*, 1955, **3**, 851.
17. M. Rejzek and R. A. Stockman, *Tetrahedron Lett.*, 2002, **43**, 6505.
18. J. F. Bunnett and S. Sridharan, *J. Org. Chem.*, 1979, **44**, 1458.
19. *Drugs of the Future*, 1996, **21**, 1105.
20. Clayden, *Organic Chemistry*, page 431.

11 General Strategy A: Choosing a Disconnection

This is the first of four *General Strategy* chapters in which we discuss important points that apply to the whole of synthetic design rather than one particular area. This chapter concerns general principles to help you choose one C–C disconnection rather than another. Even a simple molecule like the alcohol **1**, introduced in chapter 1 as a component of the elm bark beetle pheromone, can be disconnected at any of the five marked bonds.

1: component of the elm bark beetle pheromone

Greatest Simplification

Only one of the five bonds **1a** is a good choice and for two reasons. We aim to achieve the greatest simplification in our disconnections so that we get back quickly to simple starting materials. This makes the synthesis as short as possible. So we disconnect bonds that are:

(a) Towards the middle of the molecule. This breaks the molecule into two reasonably equal parts and is much better than simply lopping one atom off the end.

(b) At a *branchpoint* in the molecule: this is more likely to give simple straight chain starting materials. Here we get the aldehyde **2** and the Grignard reagent **3** coming from the straight chain halide **4**. Both **2** and **4** are commercially available.

The synthesis is a one-step process derived from the chemistry we were discussing in the last chapter.[1]

Organic Synthesis: The Disconnection Approach. Second Edition   Stuart Warren and Paul Wyatt
© 2008 John Wiley & Sons, Ltd

We can extend these guidelines when we realise that a junction between a ring and a chain and, even more, a junction between two rings, is always at a branchpoint. The series of drugs based on bicyclic structure **5** has an excellent disconnection between the two rings.

The Grignard reagent **7** is made from the halide that comes eventually from phenol by chlorination and methylation.[2] We shall discuss the synthesis of ketones like **6** in chapter 19.

Symmetry

We saw symmetry put to good use at the end of the last chapter and it may well help if we can do two identical disconnections at once. The symmetrical tertiary alcohol **10** can be made from two molecules of a Grignard reagent **11** and one of ethyl acetate. Then back to the alcohol **12** by FGI and a disconnection at the branchpoint gives the starting materials.

This synthesis was carried out by Grignard himself.[3] The bromide **16** was made from the alcohol **12** with PBr$_3$—a good reagent when an S$_N$2 displacement is needed.

Recognisable Starting Materials

A very practical guideline is to look for starting materials you can buy. It is obviously impossible to give a list of such compounds (though we've done our best with some simple compounds) or for any individual chemist to know what's available. In addition, what's available varies from year to year and sometimes from week to week. Then the requirements of research chemists, who often need only a gram of something, development chemists, who need kilograms, and production chemists who need tonnes are very different indeed, especially in the price they are prepared to pay. The solution is to use suppliers' catalogues. They are free on request from the main suppliers such as Aldrich or Fluka.

Still using symmetry, the tertiary alcohol **17** needs a butyl nucleophile and a methacrylate ester.[4] You can buy a railway tanker full of BuLi if you want and methacrylate esters are used in enormous quantities to make polymers. You won't always be as lucky as this.

In less favourable circumstances, look for starting materials that can easily be made. The hydroxyaldehyde **20** was an intermediate in Büchi's synthesis[5] of the natural product nuciferal **21**.

The tertiary alcohol is an obvious place to disconnect. Rejecting the poor disconnection of one carbon atom **20a**, we have a choice between **20b** and **20c** giving one of two ketones **23** or **24** and Grignard reagents made from one of two halides **22** or **25**. We can easily make **22** and **24** by halogenation or Friedel-Crafts acylation of toluene. But what about **23** and **25**?

We have not met methods to make **23** and, though it could be made, there is a serious chemose-lectivity problem in getting the Grignard reagent to attack the less reactive ketone in the presence of the aldehyde. Büchi preferred route **c** as he knew how to make the protected version **27** of **25** that will be needed for the Grignard reagent. We know this too as it was compound **32** in chapter 10. The Grignard reagent from **27** was combined with the ketone **24** to give a protected version of the intermediate **20**. Büchi preferred to keep the acetal in the remaining steps of the synthesis.

Returning to a series of compounds from the last chapter, Bunnett and Sridharan made one of them **29** by a different route. They went back to the alcohol as before but then disconnected the two methyl groups **30**. One reason was that it is difficult to add a MeO group to a benzene ring, but the main reason was that the methyl ester **31** is readily available.

The synthesis is straightforward but no yields are given in the paper.[6]

Available Compounds

A small selection of commonly used compounds but the Aldrich catalogue has over 34,000 entries.

Straight Chain Compounds: C_1 to about C_{10} and more in many cases
    Alcohols, alkyl halides, acids, aldehydes, amines, nitriles, ketones.

Branched Chain Compounds: as above based on these skeletons (and others):

i-propyl   *t*-butyl   *i*-butyl   *t*-amyl   1-X-3-Me-butane   2-ethylhexyl

Cyclic Compounds: C_4 to C_{10} and others:
    Ketones, alcohols, alkenes, halides, amines.

Aromatic Compounds: Very considerable variety—see catalogues.

Heterocyclic Compounds: Saturated and unsaturated in great variety.

Monomers for Polymers: Butadiene, isoprene, styrene
    Acrylates, methacrylates, unsaturated nitriles, chlorides and aldehydes.

Summary of Guidelines for Good Disconnections

1. Make the synthesis as short as possible.
2. Use only disconnections corresponding to known reliable reactions.
3. Disconnect structural C–X bonds first and try to use two-group disconnections.
4. Disconnect C–C bonds using the FGs in the molecule.
    (a) Aim for the greatest simplification. If possible
        –disconnect near the middle of the molecule
        –disconnect at a branch point
        –disconnect rings from chains
    (b) Use symmetry (if any).
5. Use FGI to make disconnections easier.
6. Disconnect back to available starting materials or ones that can easily be made.

Only some of these guidelines may apply to any given target molecule and they may well contradict each other. Developing judgement in choosing good disconnections can come only with practice. There are many different approaches to any reasonably complicated target and no 'right' answer.

References

1. R. M. Einterz, J. W. Ponder and R. S. Lenox, *J. Chem. Ed.*, 1977, **54**, 382.
2. G. M. Badger, H. C. Carrington and J. A. Hendry, *Brit. Pat.*, 1946, 576,962; *Chem. Abstr.*, 1948, **42**, 3782g.
3. V. Grignard, *Ann. Chim, (Paris)*, 1901 (7), **24**, 475.
4. P. J. Pearce, D. H. Richards and N. F. Scilly, *J. Chem. Soc., Chem. Commun.*, 1970, 1160.
5. G. Büchi and H. Wüest, *J. Org. Chem.*, 1969, **34**, 1122.
6. J. F. Bunnett and S. Sridharan, *J. Org. Chem.*, 1979, **44**, 1458.

12 Strategy V: Stereoselectivity A

Background Needed for this Chapter Reference to Clayden, *Organic Chemistry:*
Chapter 16: Stereochemistry.

The biological properties of organic molecules depend on their stereochemistry. This is true for drugs, insecticides and insect pheromones, plant growth regulators, perfumery and flavouring compounds, as indeed for all compounds having biological activity. The *cis*-hydroxyaldehyde **1** has a strong and pleasant smell and is used in lily of the valley perfumes, whereas the *trans* isomer **2** is virtually odourless. Notice that these are diastereoisomers: the compounds are achiral. Any useful synthesis must give pure **1**, not a mixture of **1** with the more stable diequatorial **2**–at equilibrium there is 92% of **2** and only 8% of **1**.

3; chiral centres
in multistriatin

The elm bark beetle pheromone multistriatin **3** is a more complicated example. You may recall from chapter 1 that a single isomer alone attracts the beetle. Making the right diastereoisomer by stereoselective synthesis is not enough. The compound must be a single enantiomer too. In this chapter we consider making the right diastereoisomer of compounds with several chiral centres and first address the question of making single enantiomers. This is only a brief discussion. You are referred to *Strategy and Control*[1] for a much more detailed analysis and to Clayden *Organic Chemistry* for the background.[2]

Enantiomerically Pure Compounds

We shall discuss two strategies in the making of single enantiomers. Either we can resolve a racemic compound somewhere in the course of the synthesis or we can use a single enantiomer as starting material. Other strategies are discussed in detail in *Strategy and Control*.

Resolution

Enantiomers cannot be separated by the normal processes of purification: crystallisation, distillation or chromatography. But diastereoisomers can. Resolution involves using an enantiomerically

Organic Synthesis: The Disconnection Approach. Second Edition   Stuart Warren and Paul Wyatt
© 2008 John Wiley & Sons, Ltd

pure 'resolving agent' to convert our racemic compound into a mixture of diastereoisomers that can be separated by these processes. When Cram[3] wanted to study the stereochemistry of elimination reactions he needed a strong enantiomerically pure base that would not substitute. In other words an asymmetric version of LDA. He chose **4**, obviously obtained from **5** and BuLi. The usual FGI and C–N cleavage **6** led back to the acid chloride **7** of available pivalic acid (*t*-BuCO$_2$H) and the amine **8**.

He prepared amine **8** by a kind of reductive amination of the ketone **9** via the *N*-formyl amine **10** and made it enantiomerically pure by resolution with malic acid **11**–a cheap enantiomerically pure compound.[4]

This would not be necessary nowadays as the preparation and resolution of **8** is an undergraduate experiment.[5] A more normal reductive amination gives racemic **8** and crystallisation of the tartrate salt **12** from methanol gives enantiomerically pure (+)-(*R*)-**8** after neutralisation. In fact this nearly perfect resolution gives both enantiomers of **8**. One tartrate salt crystallises out from MeOH and the other remains in solution. The salts are diastereoisomers and have different physical properties. Since no covalent bond is formed in making the salt **12**, simple neutralisation with NaOH gives pure amine **8** and the tartaric acid remains in solution as its sodium salt.

Cram finished his synthesis by making and reducing the amide **6**. Both steps go in excellent yield and, more importantly, without any racemisation as the chiral centre is not involved in either step. These principles are involved in all classical resolutions.

Enantiomerically Pure Starting Materials

There are very many enantiomerically pure starting materials available cheaply from nature. The amino acids are varied in structure and the hydroxyacids such as malic acid **11** and lactic acid **13** provide another resource. We shall give just one example of this kind of synthesis. Ethyl lactate **14** can be converted into the mesylate (a leaving group like tosylate) **15** and then reduced to the primary alcohol **16** with alane made from LiAlH₄ and concentrated H_2SO_4. This is not isolated but gives the epoxide **17** on treatment with base. The chiral centre is specifically inverted in the intramolecular S_N2 reaction.[6]

Stereospecific and Stereoselective Reactions

Stereospecific Reactions

Whether you are dealing with enantiomerically pure or racemic compounds, once the first chiral centre (or centres) is in place, new chiral centres must be introduced. Stereo*specific* reactions give specific and predictable stereochemical outcomes because the mechanism of the reaction demands this. The formation of **17** from **16** had to give that enantiomer as the nucleophilic oxyanion had to approach the chiral centre from the back (inversion) as all S_N2 reactions must go with inversion. Starting with enantiomerically pure materials, each enantiomer of the tosylate **18** must react in an S_N2 reaction to give an inverted acetate. One enantiomer of **18** gives one enantiomer of **19** and the other enantiomer of **18** gives the other enantiomer of **19** by stereospecific inversion.

If we are dealing with diastereoisomers the same thing applies. Compound **20** is not chiral so the question of enantiomers doesn't arise but each diastereomer of **20**, *syn* or *anti* gives a different diastereomer of **21** with inversion.

Dihydroxylation of an alkene with OsO₄ is a specifically *cis* reaction: the two OH groups add to the same side of the alkene. So *E*-**22** gives one diastereomer (*syn* as drawn) of the diol **23** while *Z*-**22** gives, by *syn* addition, a diol that can be re-drawn after rotation of a bond, as *anti*-**23**.

However, should you wish to make both *syn* and *anti*-diols from an alkene when only one isomer (*E*- or *Z*-) can be made, such as cyclopentene **25**, you need another method. Epoxidation is also a *syn*-specific method but opening the epoxide ring by an S_N2 reaction inverts one of the centres to set up an *anti* relationship. Strongly basic reagents are best avoided so acetate can be used as the nucleophile and the ester **27** can be cleaved with ammonia in methanol with attack only at the carbonyl group.

syn-**24**   **25**   **26**   **27**   anti-**24**

The table gives a list of a few stereospecific reactions but a knowledge of the mechanism of any reaction you contemplate in a synthesis is the one essential way to be sure of the stereochemical outcome.

Stereospecific Reactions

Reaction	Chemistry	Result
Substitution S_N2	R^1, R^2 …X — Nu$^\ominus$ → R^1, R^2 — Nu ; R, R …O — Nu$^\ominus$ → R, R, OH, Nu	inversion
Elimination E2	R, H, X, R — base → R, R ; R, H, X, R — base → R, R	anti-peri-planar H and X
Electrophilic addition to alkenes	R, R — RCO$_3$H → R, R, O ; R, R — OsO$_4$ → R, R, OH, OH	*cis* addition
Electrophilic addition to alkenes	R, R — Br$_2$ → R, R, Br, Br ; R, R — PhSCl → R, R, SPh, Cl	*trans* addition
Hydrogenation of alkynes and alkenes	R, R — H$_2$, Pd/C poisoned → R, H, R, H ; R^1, R^2, R^1, R^2 — H$_2$, Pd/C → R^1, H, R^2, R^1, R^2, H	*cis* addition
Rearrangements	R*, X or R*, X → Y, R*	retention at R* inversion at migration terminus retention
Reactions not involving chiral centre(s)	anything	retention

The entry 'rearrangement' may surprise you but it can be very valuable as in an alternative synthesis[7] of the amine **8**. Enantiomerically pure acid **28** is converted into the azide **29** that loses nitrogen to give a nitrene. This nitrogen atom has only six electrons and an empty orbital into which the whole side chain can migrate **30**. It does so with at least 99.6% retention of configuration to give the isocyanate **31** that picks up water to give the unstable carbamic acid **32** which loses CO_2 spontaneously to give the amine **8**. The acid **28** is not now available in enantiomerically pure form so the resolution with tartaric acid is now preferred. In any case, both enantiomers of the amine **8** are available and we would now probably use it to resolve the acid **28**.

Diastereoselective Synthesis of Multistriatin

We promised in chapter 1 that a synthesis of the elm bark beetle would appear and here it is. It has four chiral centres but one of them (marked as a hidden carbonyl group) is unimportant. Disconnecting the acetal reveals keto-diol **33**. If we make **33** it must cyclise to **3**—no other stereochemistry is possible. Further C–C disconnection with alkylation of an enolate in mind reveals symmetrical ketone **34** and a diol **35** with a leaving group (X) at one end and the two chiral centres (marked with circles) adjacent.

The leaving group will come from an alcohol so the basic skeleton is a 1,2,3-triol **36** that is nearly symmetrical and becomes symmetrical with a C–C disconnection to the symmetrical epoxide **37**. Both starting materials **34** and **37** are available and are symmetrical: we just have to make **37**. The epoxide comes from the Z-alkene **38** and that can be made by Lindlar reduction of the alkyne **39**.

The alkyne is actually available as it is easily made from acetylene and formaldehyde. Two decisions remain: how do we distinguish the three alcohols in **36** and what reagent do we use for Me⁻ in the reaction on the epoxide? Protection as the cyclic acetal **40** makes epoxidation straightforward and Me_2CuLi turned out to be the best reagent for opening the epoxide. We now have two of the OHs protected **42** but they are the wrong two!

Acetal formation is thermodynamically controlled and five-membered rings are more stable than seven-membered. So the ingenious solution was to submit **42** to acid when it rearranged by acetal exchange to **43**. Now the right OH group is unprotected and it can be transformed into the iodide **44** ready for alkylation of the lithium enolate of **34**. Treatment with acid again isomerises the acetal **45** into multistriatin **3** with loss of acetone. No attempt was made to control the centre next to the carbonyl group in **45**: cyclisation gave 85% of **3** with an equatorial methyl group[8] and only 15% of the other diastereoisomer resulting from the uncontrolled centre in **45**.

But it is important that multistriatin be made in enantiomerically pure form as well as one diastereomer. Looking back over the synthesis, the first chiral intermediate is **42** and, after some failures, reaction with the isocyanate (+)-(*R*)-**46** gave a mixture of the urethanes **47** that could be separated by crystallisation. Removal of the urethane by reduction with LiAlH$_4$ gave enantiomerically pure alcohol **42** from which enantiomerically pure (>99%) multistriatin **3** could be made by the methods above.

Stereoselective Reactions

We shall use *stereoselective* to describe reactions that have two mechanistically acceptable but stereochemically different pathways so that the molecule may *select* the more favourable—i.e. faster—pathway (kinetic control) or the more stable product (thermodynamic control). These reactions commonly involve setting up one or more new chiral centres in the presence of others.

The ketone **48** could be reduced to either alcohol **49** or **50**. The equatorial alcohol **49** is more stable and so equilibrating reducing agents like *i*-PrOH with (*i*-PrO)$_3$Al give[9] mainly **49**. But the equatorial approach **51** is kinetically favoured as the two marked axial Hs hinder approach from the other side. Large reducing agents like LiAlH(O*t*-Bu)$_3$ give[10] mostly the axial alcohol **50**.

Sometimes both diastereomers of a compound are needed and then poorly diastereoselective reactions are a boon. Both *syn* and *anti* tosylates **55** were needed to study the stereochemistry of reactions.[11] Reduction of the ketoester (see preparation in chapters 19 and 21) in two stages gave a mixture of *syn* and *anti* diols **54**, separable by column chromatography.

Each diol was selectively tosylated on the primary alcohol to give *syn* and *anti* tosylates **55** which were each treated with base [the anion of DMSO: MeS(O)CH$_2^-$]—*syn*-**55** cyclised to give the bicyclic ether **56** in good yield while *anti*-**55** fragmented to give volatile hexenal **57**.

Conformational Control in Six-Membered Rings

If a new chiral centre is formed on a saturated six-membered ring, conformational control is a possibility. We have already seen conformational effects in the reduction of ketone **48** and the same kind of arguments apply to attack on ketones by carbon nucleophiles. The alcohol **59**, needed to make an analgesic **58**, can obviously be made from the ketone **60** and that is the result of conjugate addition to cyclohexenone.[12]

Addition of Me$_2$NH to cyclohexenone followed by reaction with PhLi, without isolation of **60**, gives **59** in 60% overall yield. As you would expect, a large nucleophile such as PhLi prefers to add from the equatorial side. Notice that it adds to **60** on the *same* side as the Me$_2$N group—obviously nothing to do with steric hindrance. The Me$_2$N group merely fixes the conformation and the PhLi then adds equatorially. Acylation with the anhydride gives the drug **58**.

Axial Attack to Make a Chair

When the starting material is not a chair but a flattened chair, the first priority is to make a proper chair for the product. Strangely this means axial attack by nucleophiles on such electrophiles as cyclohexenone **62**, epoxides **65**, and bromonium ions **68**. Though the products rapidly equilibrate to all equatorial conformations **64e**, **67e** and **69e**, they are formed initially in axial **64a** or *trans*-di-axial conformations **67a** and **69a**.

Stereochemical Control in Folded Molecules

If two small rings (3-, 4- or 5-membered) are fused together (that is with two adjacent atoms common to both) they must have *cis* stereochemistry at the ring junction and a folded conformation like a half-opened book. We saw earlier that **70** had to have a *cis* ring junction: the compound *anti*-**55** that might have cyclised to the *trans* ring junction fragmented instead. This compound has a folded conformation **70a**. We shall deal with folded conformations in chapter 38 but meanwhile, notice that **70a** and **72a** show an 'outside' (the cover of the book) and an 'inside' (the pages of the book). Epoxidation of **71** goes on the outside of the folded molecule to give **72**.

Control in epoxidation of small rings by another substituent is easy to understand as the rings are virtually flat and we do not have to worry about axial and equatorial substituents. So the simple (achiral) cyclopentene **73**, with a large substituent (R = *t*-BuMe$_2$Si) on the ring, gives the *anti*-epoxide **74** because of steric hindrance. However, the free alcohol **75** epoxidises on the same face as the OH group to give **76**. The only reasonable explanation is that the OH group hydrogen bonds to the reagent and delivers it to the same face.[13]

References

1. P. Wyatt and S. Warren, *Organic Synthesis: Strategy and Control*, Wiley, Chichester, 2007.
2. Clayden *Organic Chemistry* chapters 16, 18, 32, 33 and 34.
3. D. J. Cram and F. A. A. Elhafez, *J. Am. Chem. Soc.*, 1952, **74**, 5851.
4. A. W. Ingersoll, *Org. Synth. Coll.*, 1943, **2**, 503, 506.
5. A. Ault, *J. Chem. Ed.*, 1965, **42**, 269.
6. B. T. Golding, D. R. Hall and S. Sakrikar, *J. Chem. Soc., Perkin trans 1*, 1973, 1214.
7. A. Campbell and J. Kenyon, *J. Chem. Soc.*, 1946, 25.
8. W. J. Elliott and J. Fried, *J. Org. Chem.*, 1976, **41**, 2469, 2475.
9. E. L. Eliel and R. S. Ro, *J. Am. Chem. Soc.*, 1957, **79**, 5995.
10. J. Klein, E. Dunkelblum, E. L. Eliel and Y. Senda, *Tetrahedron Lett.*, 1968, 6127; see also Wyatt and Warren, *Organic Synthesis: Strategy and Control*, chapter 21.
11. G. Kinast and L.-F. Tietze, *Chem. Ber.*, 1976, **109**, 3626.
12. M. P. Mertes, P. E. Hanna and A. A. Ramsey, *J. Med. Chem.*, 1970, **13**, 125.
13. M. Asami, *Bull. Chim. Soc. Jpn.*, 1990, **63**, 1402.

13 One Group C–C Disconnections II: Carbonyl Compounds

Background Needed for this Chapter Reference to Clayden, *Organic Chemistry:*
Chapter 9: Using Organometallic Reagents to Make C–C Bonds.
Chapter 26: Alkylation of Enolates

In chapter 10 we compared C–C disconnections with related two-group C–X disconnections, mainly at the alcohol oxidation level. In this chapter we deal more fully with carbonyl compounds, chiefly aldehydes and ketones, by two related disconnections. We start by comparing the acylation of heteroatoms by acid derivatives such as esters (a 1,1-diX disconnection **1** that can also be described as a one-group C–X disconnection) with the acylation of carbon nucleophiles and move on to compare the 1,2-diX disconnection **3** with the alkylation of enolates **6**. Here we have reversed the polarity. We mention regioselectivity—a theme we shall develop in chapter 14.

1,1-diX Disconnections: **The Corresponding C–C Disconnection:**

1,2-diX Disconnections: **The Corresponding C–C Disconnection:**

Synthesis of Aldehydes and Ketones by Acylation at Carbon

The disconnection **2a** is not useful because, as MeO$^-$ is the best leaving group from the tetrahedral intermediate **7**, the ketone **2** is formed during the reaction. The ketone is more electrophilic than the ester so it reacts again and the product is the tertiary alcohol **8**.

One solution is to use an acid chloride as an acylating agent since that is *more* electrophilic than the ketone. The problem with this approach is that we wish to combine two extremely reactive compounds and an uncontrollable reaction ensues. Successful acylation of the much less reactive, and therefore more selective, organo-copper reagents is known. Treatment of organo-lithium reagents with CuI in dry THF at−78 °C gives dialkyl copper lithiums or cuprates[1] R$_2$CuLi. These react cleanly with acid chlorides, again at low temperature, to give ketones.[2]

A simple example that also shows some chemoselectivity is the preparation of the ketones **12**; R = Et or Pr by reaction of the bromoacid chloride **11** with the appropriate dialkyl copper lithium. The bromoacid **10** is available and can be converted into a range of bromoketones by this method.[3]

Only one alkyl group is transferred from R$_2$CuLi and to avoid wasting the other R group, complexing agents can be added. Posner[4] uses a PhS group to stabilise the organo-copper reagent **14** with one *t*-Bu group that is cleanly transferred to the acid chloride. Friedel-Crafts reactions of *t*-BuCOCl are plagued with loss of CO so this is a better method.

In his synthesis of the [4.4.4]'propellane' **19**, Paquette made the diester **16** easily but wanted the *mono* methyl ketone **18**. Rather than add MeLi directly, he first hydrolysed one ester to the free acid **17** and then made the acid chloride with oxalyl chloride. Reaction with Me$_2$CuLi gave the ketone in excellent yield.[5]

Direct Formylation of Organo-Lithiums with DMF

Another method is to use a much *less* electrophilic acylating agent than an ester. This sounds crazy but DMF **20** reacts directly with organo-lithium compounds to give good yields of aldehydes

23. Now the tetrahedral intermediate **21** is stable under the reaction conditions as Me_2N^- is such a bad leaving group. The aldehyde **23** is formed only during work-up in aqueous acid **22**.

The organo-lithium reagent can be made by exchange of Li for a halide or by deprotonation. With di-iodide **24**, one iodine may be exchanged with one equivalent of BuLi and the aldehyde **25** is the product.[6] The aromatic heterocycle isothiazole **26** has its most acidic hydrogen (marked) next to sulfur and it gives one aldehyde **27** in good yield.[7]

A more reactive equivalent for ketone synthesis is a nitrile **28**. Addition of a Grignard reagent gives an intermediate **29**, stable under the reaction conditions, rather like **21**. Hydrolysis in acid solution releases the ketone **2**. The exactly analogous reagent to DMF would be a tertiary amide but these are often so unreactive as to be useless.[8]

Grignard reagents are usually better than organo-lithiums in these reactions[9] and may be even better with a catalytic amount of copper (I). A good example is the coupling of the Grignard reagent derived from **30** with the protected nitrile **31** giving an excellent yield of the ketone **32**. As a bonus, the protecting group drops off in the work-up.[10]

These reactions may be intramolecular giving five- or six-membered cyclic ketones such as the spiro (two rings with one common atom) compound **34** and the hindered cyclohexanone **36**.[11]

Please note that this whole section—indeed all of the chapter so far—relates to the C–C disconnection between the carbonyl group and whatever is joined to it **2**. The nucleophilic reagent is an organometallic derivative of Li, Cu or Mg and the electrophile is an acid chloride, a tertiary amide or a nitrile. In the next section we disconnect the C–C bond one further from the carbonyl.

Carbonyl Compounds by Alkylation of Enols

Disconnection **37** again uses the natural polarity of the carbonyl group but at the next bond **37** since we hope to use some enolate derivative **38** in an alkylation reaction. But—and it is a big but—do not think for a moment that you can make **37** just by mixing the ketone **39** with an alkyl halide and some base. The problem is that the ketone is itself electrophilic and the self-condensation by the aldol reaction (chapter 19) is generally preferred to alkylation.

We need first of all to convert the ketone **39** *completely* into some enolate derivative so that there is no ketone left for self-condensation. In this chapter we shall restrict ourselves to lithium enolates **40** and anions **42** of 1,3-dicarbonyl compounds **41**. Each of these reagents acts as the enolate anion of acetone **38**; R^2 = Me.

Lithium Enolates of Simple Carbonyl Compounds

Lithium enolates **40** are usually made with LDA (Lithium Di-isopropylAmide). We need a strong base—one strong enough to convert the ketone immediately into the lithium enolate. Butyl lithium would be strong enough but it attacks the carbonyl group as a nucleophile instead. We therefore use the BuLi to make LDA—a strong but very hindered base that usually does not attack the carbonyl group. The reagent LDA is prepared in dry THF at low temperature[12] and the ketone added by syringe also at low temperature. The lithium atom bonds to the oxygen and the amide is then in perfect position to remove the proton **43**.

If the ketone is symmetrical, as here, or can form an enolate on one side only, or if we are dealing with an ester, enolate formation and hence alkylation is unambiguous. In Corey's synthesis of cafestol,[13] an anti-inflammatory agent from coffee beans, he first alkylated ketone **44** on the only possible side and converted the product **45** into the new alkylating agent **46**.

Next he made the lithium enolate **48** from the unsaturated ester **47**—only the marked hydrogen can be removed—and alkylated this with **46**. Almost all of the skeleton of cafestol is assembled in this important step.

Enolates of 1,3-Dicarbonyl Compounds

You should appreciate that syringe techniques in scrupulously dry apparatus and solvents at $-78\,^{\circ}C$ are not the easiest. An alternative to using a very strong base is to modify the ketone so that the enolate is formed much more easily. This is done by adding an ester group **41** that has the sole function of making the enolate **42** conjugated—the negative charge is shared by both oxygen atoms. Only a relatively weak base is needed to make the enolate **42** and the usual choice is the alkoxide of the ester. Alkylation occurs on the middle carbon and the product **50** can be decarboxylated by ester hydrolysis and heating the free acid.

The hydrolysis gives the anion **52** that is protonated to give the keto-acid **53**. Often spontaneously, but always on heating, decarboxylation by a cyclic mechanism **54** gives the enol **55** of the alkylated ketone **51**.

The extra ester group is not normally added to the preformed ketone as ethyl acetoacetate **41** is available and the diester is available diethyl malonate **59**. If it is necessary to make the 1,3-dicarbonyl compound, this can be done by methods described in chapters 19 and 20. The carboxylic acid **56** can be disconnected at the branchpoint to an alkyl halide and the synthon **58** that could be realised as the anion of diethyl malonate **59** or the lithium enolate of ethyl acetate.

One published synthesis uses the malonate route.[14] Ethoxide is used as the base so that it doesn't matter if it attacks the esters as a nucleophile.

A good example of a ketone made by this strategy is used in the synthesis of terpenes. After the usual '1,2' C–C disconnection, adding the ester group to the enolate **64** gives ethyl acetoacetate **41**.

The alkylation goes well as **63** is the reactive allylic halide 'prenyl bromide' and hydrolysis and decarboxylation occur as usual.[15]

Carbonyl Compounds by Conjugate Addition

The remaining style of C–C disconnection takes us straight to conjugate addition and we are still using the natural polarity of the carbonyl group. Conjugate addition of a heteroatom to the enone **66** gives the 1,3-relationship in **65** and the same process with a carbon nucleophile gives **67**.

1,3-diX Disconnections: **The Corresponding C–C Disconnection:**

We can use either organo-lithiums or Grignard reagents as the carbon nucleophiles but we need copper (I) to ensure conjugate addition. Without Cu(I) both nucleophiles are inclined to add directly to the carbonyl group. We can use the same reagents that we used to make ketones in this chapter.

In Corey's synthesis[16] of a marine allomone, he wanted the cyclic ketone **68**. The Friedel-Crafts disconnection gives some derivative of the carboxylic acid **69** and disconnection between the branchpoints gives the unsaturated acid **70** (it doesn't matter whether this is the *E*- or *Z*- isomer as the alkene disappears).

In practice he used the *E*-unsaturated ester **71**, as that was easier to make, and added isopropyl Grignard with a CuSPh catalyst (see compound **13** above) to avoid wasting one equivalent of the Grignard. The ester product **72** cyclised to the target with polyphosphoric acid without a specific ester hydrolysis step. No doubt this works so well because it is an intramolecular reaction giving a five-membered ring.

Aromatic compounds are good enough nucleophiles to add in conjugate fashion under Friedel-Crafts conditions so that no organo-metallic reagent is needed. Benzene adds to cinnamic acid **74** with AlCl₃ as catalyst to give **73** in one step.[17]

In the synthesis of the diol **75** stereochemistry is important. The diol could be made from the keto-ester by stereoselective reduction using a suitable reducing agent (chapter 12) as the alcohol on the six-membered ring is axial.[18]

Disconnection of the ketone **76** with conjugate addition in mind could remove the vinyl group **76a** or the methyl group **76b**. There are two reasons why we prefer **a**. The addition is likely to occur from the opposite face of the molecule to the CO_2Et group and that is where we want the vinyl group. Conjugate addition to **78** might occur at the β-position but it could equally well occur at the very exposed δ-position. The starting material **77** is also the available Hagemann's ester **77**.

The vinyl Grignard reagent was used with Cu(I) catalysis and the reduction of both ester and ketone was achieved with $LiAlH_4$. The stereoselectivity was excellent and **75** could easily be separated from the minor equatorial alcohol. In the next chapter we shall revisit both the use of copper in getting regioselectivity and the stereoselectivity of such reactions.

References

1. *Vogel*, page 483.
2. *Vogel*, page 616.
3. J. A. Bajgrowicz, A. El Hallaoui, R. Jacquier, C. Rigieri and P. Viallefont, *Tetrahedron*, 1985, **41**, 1833.
4. G. H. Posner and S. E. Whitten, *Org. Synth.*, 1976, **55**, 123.
5. H. Jendralla, K. Jelich, G. de Lucca and L. A. Paquette, *J. Am. Chem. Soc.*, 1986, **108**, 3731.
6. Clayden, *Lithium*, page 124.
7. M. P. L. Caton, D. Jones, R. Slack and K. R. H. Wooldridge, *J. Chem. Soc.*, 1964, 446.
8. Kharasch and Reinmuth pages 767– 845.
9. F. J. Weiberth and S. S. Hall, *J. Org. Chem.*, 1987, **52**, 3901.
10. I. Matsuda, S. Murata and Y. Izumi, *J. Org. Chem.*, 1980, **45**, 237.
11. M. Larcheveque, A. Debal and T. Cuvigny, *J. Organomet. Chem.*, 1975, **87**, 25.
12. *Vogel*, page 603.
13. E. J. Corey, G. Wess, Y. B. Xiang and A. K. Singh, *J. Am. Chem. Soc.*, 1987, **109**, 4717.
14. E. B. Vliet, C. S. Marvel and C. M. Hsueh, *Org. Synth. Coll.*, 1943, **2**, 416.
15. J. Weichet, L. Novak, J. Stribrny and L. Blaha, *Czech Pat.*, 1964, 112,243; *Chem. Abstr.*, 1965, **62**, 13049e.
16. E. J. Corey and W. L. Seibel, *Tetrahedron Lett.*, 1986, **27**, 905.
17. P. Pfeiffer and H. L. de Waal, *Liebig's Ann. Chem.*, 1935, **520**, 185.
18. T. Kametani and H. Nemoto, *Tetrahedron Lett.*, 1979, 3309.

14 Strategy VI: Regioselectivity

Background Needed for this Chapter References to Clayden, *Organic Chemistry:*
Chapter 10: Conjugate Addition, Chapter 26: The Alkylation of Enolates

Chapter 5 dealt with *chemo*selectivity: how to react one functional group rather than another. Now we must face a more subtle and demanding problem: how to react one specific part of a single functional group and no other. This is *regio*selectivity. We have already seen that anions of phenols **2** are alkylated at oxygen to give ethers **3** while enolate anions **5** are alkylated at carbon to form a new C–C bond **6**.

By and large in those two cases what you get is what you want. We shall look at two important aspects of regioselectivity where we may want either result. We should like to be able to alkylate unsymmetrical ketones **8** on one side or the other to give **7** or **9**. We should like to add nucleophiles to enones **11** either directly at the carbonyl group to give **12** or in conjugate fashion, as we have been discussing in the last chapter, to give **10**.

The Regioselective Alkylation of Ketones

In the last chapter we used two *specific* enol equivalents for alkylation reactions: lithium enolates and 1,3-dicarbonyl compounds. Both will help us to solve the regioselectivity problem in the

Organic Synthesis: The Disconnection Approach. Second Edition   Stuart Warren and Paul Wyatt
© 2008 John Wiley & Sons, Ltd

alkylation of unsymmetrical ketones. Suppose we want to make **13**. At first sight it appears that we must alkylate an unsymmetrical ketone on the more substituted side. But, if we remove the benzyl group and add our activating CO_2Et group to give **14** it is clear that we can make this by another alkylation and the activating group will promote both.

Benzyl is the more reactive bromide so it makes sense to add it last since making the quaternary carbon will be difficult. This was the order followed in the published synthesis.[1]

If we had wanted the isomeric ketone **19** with the benzyl group on the other side of the carbonyl, we could use a property of lithium enolates we have kept secret until now. Enolate formation with LDA goes on the less hindered alkyl side chain, particularly if it is a methyl group. We could start with acetone **16**, alkylate on one side with PrBr to give **17** and then treat again with LDA. Now kinetic selectivity gives the lithium enolate **18** (not isolated) and benzylation must give **19**. This regioselectivity might be clear if you look at the intramolecular mechanism for proton removal in chapter 13, diagram **43**. An alkyl group on the carbon atom being deprotonated would be sterically hindering.

This looks very clean but a careful study[2] of the closely related ketone **20** shows that the ratio of less substituted **21** to more substituted enolate **23** is 87:13 so the product **22** is inevitably contaminated with **24**.

Regioselectivity is better when the contrast is between secondary and tertiary centres as with cyclic ketone **25**. The less substituted lithium enolate **26** is formed almost exclusively (99:1) in dimethoxymethane.[3]

It is however possible to make such compounds in good yield from 1,3-dicarbonyl starting materials. Another isomer of **13** and **19** is the branched ketone **28**. Disconnecting by a method from chapter 13, we can use the acid derivative **29** which could be made from malonate **31** by two alkylations via **30**.

The published synthesis uses a cadmium reagent but we should rather use copper nowadays.[4] Double alkylation of malonate, again adding the benzyl group last, gives **33**. Hydrolysis and decarboxylation releases the free acid **34** which is easily converted into its acid chloride and then with Pr_2Cd, or perhaps better Pr_2CuLi, into the target molecule **28**.

Regioselectivity in Nucleophilic Addition to Enones

The problem of getting direct (1,2-) or conjugate (1,4- or Michael) addition to α,β-unsaturated compounds such as enones **11** can be solved without finding abstruse strategies by choice of reagents.

The general principles are:

1. The conjugate addition product **10** is thermodynamically favoured as the weak C=C bond has been lost but the strong C=C bond retained. The direct addition product **12** is kinetically favoured.
2. Direct addition is more easily reversed than conjugate addition. So the more stable the nucleophile, the more reversible 1,2-addition becomes and the more 1,4-addition predominates.

3. The C=O site is the 'harder' and the site of conjugate addition the 'softer' electrophilic site so more basic nucleophiles tend to do 1,2-addition and less basic nucleophiles tend to prefer 1,4-addition.

We saw in chapter 6 that the more electrophilic the α,β-unsaturated compound, the more likely it is to do direct addition with heteroatoms and the same is true for carbon nucleophiles. Grignard reagents normally add direct to α,β-unsaturated aldehydes[5] such as **35** but may add in conjugate fashion to α,β-unsaturated esters,[6] particularly if they have a large esterifying group, such as the *sec*-butyl group in **37**.

Among nucleophiles we saw again in chapter 6 that the softer, less basic nucleophiles such as sulfur are good at conjugate addition, hydride reducing agents and RO⁻ are better at direct addition, while amines are somewhere in the middle. This means[7] that α,β-unsaturated aldehydes, ketones and esters may all be reduced at the carbonyl group with $NaBH_4$ (for aldehydes and ketones) or $LiAlH_4$ for esters such as **40**. Catalytic hydrogenation however is not an ionic reaction and simply reduces weak bonds[8] such as C=C and not C=O.

An example is the preparation of the unsaturated alcohol **42** whose benzyl ether was needed for a Diels-Alder reaction. Reduction of the diene ester **43** with $LiAlH_4$ gave 85% of the alcohol. The starting material **43** is easily made[9] by methods discussed in chapters 15 and 19.

Carbon Nucleophiles in Conjugate Addition

The very basic and aggressive nucleophilic organo-lithiums tend to do direct addition to all α,β-unsaturated carbonyl compounds. The rather less basic Grignard reagents may add in either sense, as we saw for compounds **35** and **37**. We saw in the last chapter how Cu(I) is the key to persuading either RLi or RMgBr to add in a conjugate fashion.[10] If we react **11** with a Grignard reagent with Cu(I) catalysis we get **44** as the product. With R_2CuLi the initial product is actually the lithium enolate **45** giving us the opportunity to add an electrophile to make **46**. Of course, if the electrophile is a proton, the product is still **44**.

As we saw in chapter 13, addition of a cuprate to a cyclic enone such as **47** followed by trapping with an electrophile gives *anti* stereochemistry **48**.

So when chemists at Dortmund wished to make the *syn* compound **49**, they chose to add diallyl copper lithium to the enone **50** with one side chain already in place.[11] This gave the lithium enolate **51** and protonation gave the *syn* compound **49**. The choice of acid was important: phenols were good and **52** was the best.

Thermodynamic control dominates when a *cis* ring junction is preferred, as between the flat five-membered and the six-membered ring in **55**. Reversible protonation of the lithium enolate **54** occurs on the same face as the methyl group on the *exo* face (chapter 12). The molecule prefers a folded conformation.

Stereochemistry may be created during the addition step as in the formation of **57** and **59**. Cuprate addition to **56** gives[12] a 98:2 ratio of *anti:syn* products **57**. The addition of PhCu, prepared from PhLi and Cu_2I_2 (and perhaps Ph_2CuLi) to **58** gives[13] a 96:4 *anti:syn* ratio of **59**. In both cases the cuprate has added from the opposite face to the substituent but the stereochemistry of **57** is probably dominated by axial attack giving a chair product (chapter 12) while **59** has both substituents equatorial.

But most of the time we are chiefly concerned with getting the carbon nucleophile to add 1,4. The synthesis of the unsaturated ketone **62** makes two more points. While we cannot do S_N2 reactions on vinyl halides, they make good organo-lithium and copper reagents and, in the addition to cyclohexenone, propenyl bromide both forms the cuprate and adds with retention at the alkene.[14] This leads us on to the next chapter.

References

1. J. Cason, *Chem. Rev.*, 1947, **40**, 15; D. A. Shirley, *Org. React.*, 1954, **8**, 28.
2. C. L. Liotta and T. C. Caruso, *Tetrahedron Lett.*, 1985, **26**, 1599.
3. H. O. House, M. Gall and H. D. Olmstead, *J. Org. Chem.*, 1971, **36**, 2361; M. Gall and H. O. House, *Org. Synth.*, 1972, **52**, 39.
4. L. Clarke, *J. Am. Chem. Soc.*, 1911, **33**, 529; W. B. Renfrew, *Ibid.*, 1944, **66**, 144; C. S. Marvel and F. D. Hager, *Org. Synth. Coll.*, 1932, **1**, 248; J. R. Johnson and F. D. Hager, *Ibid.*, 351.
5. E. Urion, *Comp. Rend.*, 1932, **194**, 2311; *Chem. Abstr.*, 1932, **26**, 5079.
6. J. Munch-Petersen, *Org. Syn. Coll.*, 1973, **5**, 762.
7. W. G. Brown, *Org. React.*, 1951, **6**, 469.
8. House pages 1–34.
9. J. Auerbach and S. M. Weinreb, *J. Org. Chem.*, 1975, **40**, 3311.
10. G. H. Posner, *Org. React.*, 1972, **19**, 1; 1975, **22**, 253; H. O. House, *Acc. Chem. Res.*, 1976, **9**, 59; J. F. Normant, *Synthesis*, 1972, 63.
11. N. Krause and S. Ebert, *Eur. J. Chem.*, 2001, 3837.
12. H. O. House and W. F. Fischer, *J. Org. Chem.*, 1968, **33**, 949.
13. N. T. Luong-Thi and H. Riviere, *Compt. Rend.*, 1968, **267**, 776.
14. C. P. Casey and R. A. Boggs, *Tetrahedron Lett.*, 1971, 2455.

15 Alkene Synthesis

Background Needed for this Chapter Reference to Clayden, *Organic Chemistry:*
Chapter 19: Elimination Reactions; Chapter 31: Controlling the Geometry of Alkenes

Synthesis of Alkenes by Elimination Reactions

Alkenes can be made by the dehydration of alcohols **2**, usually under acidic conditions, the alcohol being assembled by the usual methods. This route is particularly good for cyclic alkenes **3** and those made from tertiary and/or benzylic alcohols as the E1 mechanism works well then. The same alkene is formed[1] from **2** regardless of which side eliminates but **4** gives a 76% yield of an 80:20 mixture of **5** and **6**.

Acids must be fairly strong for this job and must have a non-nucleophilic counterion to avoid substitution. Popular ones are $KHSO_4$ and TsOH (crystalline and easier to handle than H_2SO_4 or H_3PO_4) and the less acidic $POCl_3$ in pyridine. Little control is found over the position or geometry of the alkene though in many simple cases, such as **2**, this doesn't matter. Notice however that if R = Alkyl, very little exocyclic alkene, if any, is among the product.

When Zimmermann and Keck[2] wished to study the photochemistry of a series of alkenes of the general structure **7** they could have put the OH group at either end of the double bond but they chose the branchpoint **8** because dehydration of the tertiary benzylic alcohol should be very easy and there is no ambiguity in the position of the alkene whatever R may be. They used the Grignard method and dehydrated **8** with $POCl_3$ in pyridine.

Eliminations on alkyl halides follow essentially the same strategy except that the reaction is now done by the E2 mechanism with a strong hindered base to avoid S_N2 reactions. This

---

Organic Synthesis: The Disconnection Approach. Second Edition   Stuart Warren and Paul Wyatt
© 2008 John Wiley & Sons, Ltd

approach is good for terminal alkenes **10** as the elimination is successful on primary halides. The alcohol **12** can be made by any method (chapter 10).

So a typical synthesis might involve treating the alcohol **10** with PBr$_3$ to make the bromide and eliminating with *t*-BuOK. There is again no ambiguity in the position of the alkene.

Dienes can be made by this elimination strategy if vinyl Grignards are used as the vinyl group blocks dehydration in that direction and makes the cation intermediate in the E1 reaction allylic. An interesting example[3] is the four-membered ring compound **13**, disconnected via the allylic alcohol **14** to cyclobutanone **15**.

Cyclobutanone **15** is available, and also very electrophilic, so addition of the vinyl Grignard and dehydration with the rather unusual reagent iodine gave the diene **13**. This diene will be used in a Diels-Alder reaction in chapter 17.

Alkene Synthesis by the Wittig Reaction

The most important method of alkene synthesis is now the Wittig reaction[4] which gives full control over the position of the double bond and some control over its geometry. A phosphine, usually triphenyl phosphine Ph$_3$P, reacts with an alkyl halide in an S$_N$2 reaction to give a phosphonium salt **18**. Treatment with base, often BuLi, gives the phosphonium ylid **19**. An *ylid* is a species with positive and negative charges on adjacent atoms. Reaction with an aldehyde gives the alkene, usually the Z-alkene **20** if R^1 is an alkyl group, and triphenylphosphine oxide **21**.

The mechanism for the formation of the alkene is open to discussion especially as there is no agreement on the source of the stereoselectivity.[5] We suggest that the carbanion end of the ylid adds to the aldehyde **22** and the 'betaine' then cyclises **23** to the four-membered ring which fragments to give the products **24**. There is no doubt about intermediate **24** nor that its decomposition must be stereospecific. So the Z-alkene **20** is formed from the *cis* oxaphosphetane **24**.

As the Wittig reaction forms both π and σ-bonds, the disconnection is right across the middle of the alkene giving a choice of starting materials. So with the *exo*-cyclic alkene **26**, very difficult to make by elimination methods, we could use formaldehyde or cyclohexanone as the carbonyl component with either phosphonium salt **25** or **28**. It is a matter of personal choice whether you draw the ylid, the phosphonium salt or the alkyl halide at this stage.

Wittig did this synthesis[6] with the iodide **29**, which he made himself, to give **26** in low yield (46%) but higher yields are routinely obtained nowadays: Vogel reports[7] 64% from the commercially available bromide **30** using the sodium salt of DMSO as base.

Trisubstituted alkenes **32** are no trouble as either a secondary halide **35** or a ketone can be used. As both **33** and **35** are available we choose them.

The synthesis is straightforward but does produce a mixture of geometrical isomers.[8] This is another case where dehydration, even of the tertiary alcohol **36**, would probably give a mixture of positional (and geometrical) isomers.

Wittig Reactions with Stabilised Ylids

The unstable aspect of the ylid is the carbanion: phosphonium salts are stable compounds so any substituent that stabilises the anion also stabilises the ylid and this reverses the stereoselectivity to favour the *E*-alkene. Even benzylic ylids give *E*-alkenes as in the reaction[9] with the anthracene **37** that gives a good yield of crystalline **38** having a coupling constant between the two marked Hs of 17 Hz. One possible explanation is that the formation of the betaine or oxaphosphetane is reversible if the ylid is stabilised and only the faster of the two eliminations occurs to give the *E*-alkene.

37   *E*-38; 74%yield

Applications of the Wittig Reaction

An excellent application of the distinction between stabilised and unstabilised ylids is in the synthesis of leukotriene antagonists.[10] The intermediate **39** (R is a saturated alkyl group of 6, 11 or 16 carbon atoms) was needed and disconnection of the *Z*-alkene with a normal Wittig reaction in mind followed by removal of the epoxide exposed a second alkene with the *E* configuration that could be made from the aldehyde **43** and the stabilised ylid **42**.

This ylid is so stable that it is commercially available and reacts cleanly with **43** to give only *E*-**41**. Epoxidation under alkaline conditions gives the *trans* epoxide **40** and a normal Wittig on an unstabilised ylid gives **39**: the yield depends on R.

When the substituent becomes very anion-stabilising, as in **42**, the ylid may not react with ketones and anions of phosphonate esters are usually preferred in the Horner-Wadsworth-Emmons (HWE) variant.[11] The reagent triethyl phosphonoacetate **46** is made by combining a phosphite $(EtO)_3P$ instead of a phosphine, with ethyl bromoacetate. Displacement of bromide **44** gives a phosphonium ion that is dealkylated by bromide **45**.

Barrett used the reaction at the start of his synthesis of an antibiotic.[12] The HWE reaction with the enal **47** gives the diene ester **48** and by reduction with DIBAL, the dienol **49**.

The 'optical brightener' Palanil **50**—it makes your clothes look 'whiter than white' and your T-shirts fluoresce in UV light—can be disconnected by the Wittig strategy to two molecules of the phosphonium ylid **51** and one of the dialdehyde **52**. The availability of the dialdehyde, used in the manufacture of terylene, makes this route preferable to the alternative.

As the ylid **51** is stabilised by the nitrile as well as the benzene ring, the phosphonate ester **54** is preferred in the manufacture and the reaction is strongly *trans* selective.[13] The by-product is the anion of dimethyl phosphate **55** which is water-soluble and very easy to separate from the product **50**. By contrast, triphenylphosphine oxide is insoluble in water and can be difficult to separate from the alkene.

Many insect pheromones are derivatives of simple alkenes. Disparlure **56**, an attractant for the gypsy moth, is an epoxide derived by stereospecific epoxidation from the Z-alkene **57**. As neither substituent is anion-stabilising, a simple Wittig should give the right geometry.

The synthesis was carried out this way,[14] though no doubt the alternative combination would also work well. The synthetic material is as attractive to the moth as the natural pheromone.

Dienes by the Wittig Reaction

Conjugated dienes are needed for the Diels-Alder reaction (chapter 17) and Wittig disconnection **61** reveals that the choices here are more important. The easily prepared enals **62** would react with an unstabilised ylid **63** to give a Z-alkene but the conjugated allylic ylid **60** might give the E-alkene.

The mono-substituted butadiene **66** was needed with a *trans* alkene in the middle. So disconnection **61a** looks good and the allyl phosphonium salt **65** did indeed give *E*-**66** though in poor yield. These low molecular weight hydrocarbons are difficult to isolate as they are volatile.

Diaryl butadienes **69** can be made by method **61b** as the ylid from **68** is conjugated and will give one E-alkene while the other comes from an aldol condensation used to make the enal.[15] (chapter 19).

Though the Wittig is the most important, there are many other ways to make alkenes using a variety of elements in the periodic table keeping the same disconnection.[16]

References

1. *Vogel*, page 491.
2. H. E. Zimmermann, T. P. Gannett and G. E. Keck, *J. Org. Chem.*, 1979, **44**, 1982.
3. R. P. Thummel, *J. Am. Chem. Soc.*, 1976, **98**, 628,
4. A. Maercker, *Org. React.*, 1965, **14**, 270;   B. E. Maryanoff and A. B. Reitz, *Chem. Rev.*, 1989, **89**, 863.
5. E. Vedejs, *J. Org. Chem.*, 2004, **69**, 5159;   R. Robiette, J. Richardson, V. K. Aggarwal and J. N. Harvey, *J. Am. Chem. Soc.*, 2006, **128**, 2394.
6. G. Wittig and U. Schöllkopf, *Org. Synth. Coll.*, 1973, **5**, 751.
7. *Vogel*, page 498.

8. C. F. Hauser, T. W. Brooks, M. L. Miles, M. A. Raymond and G. B. Butler, *J. Org. Chem.*, 1963, **28**, 372.

9. E. F. Silversmith, *J. Chem. Ed.*, 1986, **63**, 645; Vogel, page 500.

10. T. W. Ku, M. E. McCarthy, B. M. Weichman and J. G. Gleason, *J. Med. Chem.*, 1985, **28**, 1847.

11. W. S. Wadsworth and W. D. Emmons, *J. Am. Chem. Soc.*, 1961, **83**, 1733; *Org. Synth. Coll.*, 1973, **5**, 547.

12. A. G. M. Barrett and G. J. Tustin, *J. Chem. Soc., Chem. Commun.*, 1995, 355.

13. H. Pommer and A. Nürrenbach, *Pure Appl. Chem.*, 1975, **43**, 527; *Angew. Chem., Int. Ed. Engl.*, 1977, **16**, 423.

14. C. A. Henrick, *Tetrahedron*, 1977, **33**, 1845; B. A. Bierl, M. Beroza and C. W. Collier, *Science*, 1970, **170**, 87.

15. R. N. McDonald and T. W. Campbell, *J. Org. Chem.*, 1959, **24**, 1969.

16. S. E. Kelly in *Comp. Org. Synth.*, volume 1, 1999, page 729.

16 Strategy VII: Use of Acetylenes (Alkynes)

Background Needed for this Chapter Reference to Clayden, *Organic Chemistry:* Chapter 9: Using Organometallic Reagents to Make C–C Bonds.

This strategy chapter is rather different. We shall look at one class of starting material—alkynes or acetylenes—and see what special jobs they can do in synthesis. In particular, we shall see how they can solve some problems we have already met. Acetylene itself **3** is readily available and its first important property is that protons on triple bonds are much more acidic than most CH protons. Acetylene forms a genuine anion **4** with sodium in liquid ammonia, a lithium derivative **1** with BuLi and a Grignard reagent **2** by reaction with a simple alkyl Grignard such as EtMgX.

These derivatives react with the type of carbon electrophiles we have already met such as alkyl halides, aldehydes and ketones, and epoxides to give **5**, **6** and **7** respectively.

Oblivon **8** is clearly an adduct of acetylene and a ketone **9** and the synthesis is trivial. This example and the next are from the patent literature[1] so we are guessing at the details and have no yields.

These products still have an acidic hydrogen on the triple bond so they can react again with base and an electrophile. Surfynol **10** is clearly an acetylene diadduct with the same ketone **11** being used twice.[2]

Organic Synthesis: The Disconnection Approach. Second Edition   Stuart Warren and Paul Wyatt
© 2008 John Wiley & Sons, Ltd

This time a second treatment with base will be needed to make the second anion. As the OH proton in **12** is more acidic (pK_a ~ 16) than the alkyne proton (pK_a ~ 25), two molecules of base will be needed and the most reactive 'anion' (alkyne) reacts first.

The Reduction of Alkynes to Alkenes

Otherwise few of these acetylene adducts are important in their own right but they are valuable intermediates because disubstituted acetylenes **15** can be reduced at will to either *E*- or *Z*-alkenes by different reducing agents. Catalytic hydrogenation using the Lindlar catalyst, to stop reduction to the alkane, adds a molecule of hydrogen to one side of the triple bond to give *Z*-**14**. Addition of solvated electrons, formed when sodium metal dissolves in liquid ammonia, may give the dianion **16** with the two negative charges in sp^2 orbitals as far from each other as possible, and certainly gives *E*-**14** on protonation with a weak base such as *t*-BuOH.

We saw the *cis*-selective reduction when we made compound **30** in chapter 12: the starting material *cis*-butenediol **18** is readily available as it is made by the Reppe process.[3] There is a fascinating story of the reluctant Reppe not divulging the details to the allies after the second world war.[4]

An unsymmetrical example is the allylic halide **19** needed for the synthesis of *cis* jasmone. Obvious disconnections take us back to **21** and a simple three-component synthesis.

As the alcohol **22** was available it was simply a case of putting in the ethyl group. Of course the alternative order of events might be better. Since the reduction to the *cis*-alkene can be done either before or after incorporation into the *cis*-jasmone skeleton, **22** was also transformed[5] into the propargylic bromide **24**.

The *trans* acetate **25** is the pheromone used to trap pea moths.[6] Changing the *E*-alkene into the acetylene allows the disconnection next to what was the double bond. The only problem is how to make a mono-bromide from the symmetrical diol **30**.

Experiments showed that protecting one end of the diol **30** as the THP derivative **31** (chapter 9) gave good yields and the monobromo compound **32** was used to alkylate acetylene with the methyl group added later. Reduction and deprotection gave the *E*-alcohol **26** and acetylation gave the pheromone **25**.

The Synthesis of Dienes

In chapter 15 we saw that dienes could be made by the Wittig reaction and also by the addition of vinyl lithiums or Grignard reagents to ketones followed by dehydration of the allylic alcohol product. Derivatives of acetylenes can do the same job. The first disconnection is the same but a reagent for the synthon **40** replaces the vinyl metal derivative.

The published synthesis[7] of **36** uses the sodium salt and the reduction goes in excellent yield. Only the dehydration with KHSO$_4$ is poor.

Ketones by Hydration of Acetylenes

A rather different reaction of acetylenes is the addition of water, usually catalysed by Hg(II), to give ketones. Terminal acetylenes **41** reliably give methyl ketones **44** as the intermediate vinyl cation **42** is secondary. Water adds to the cation **42** to give the enol **43** which equilibrates to the ketone **44** with the loss of $Hg(OAc)_2$.

Symmetrical acetylenes can also be hydrated to one ketone as the two possibilities are the same. An intriguing example is the hydration of the diol **45** that presumably gives the ketone **46**. This is not isolated as, under the conditions of the reaction, formation of the cyclic ether **47** is faster than the hydration.[8]

The very unsymmetrical acetylene **48**, with a ketone on one side of the alkyne and a *cis*-alkene on the other, hydrates completely regioselectively[9] to the diketone **49**.

The involvement of the ketone in the intermediate **50** shows why water adds to one end of the alkyne and not the other. We have already seen above how the *cis* double bond in molecules of this sort can be made from alkynes and so both main uses of alkynes appear here.

In fact, the diketone **49** was not isolated but was cyclised with dilute aqueous base to give cis-jasmone **52** in excellent yield. We shall be exploring reactions of this sort in chapters 18–28.

An Alkyne-Containing Anti-AIDS Drug

Merck's reverse transcriptase inhibitor efavirenz **53** is one of a new generation of anti-AIDS drugs.[10] Disconnection of two structural C–O bonds reveals **54** that is clearly the adduct of an acetylene **56** and the ketone **55**. The question is, how do we make **56**?

We have not yet met three-membered rings but cyclisation of carbon nucleophiles onto CH$_2$s with a leaving group (X in **57**) works well. Now the question is: how do we make the pentynol **58**?

The answer is an exaggerated form of starting material strategy. We need a five-carbon compound with some oxygenation and some unsaturation. There is one exceptionally cheap and abundant compound that fits the bill: furfural **59**. When breakfast cereals are made, furfural is an abundant by-product and the Quaker Oats company have patents on isolating it but chemists can do so from corn cobs by a simple recipe that gives 165–200 g furfural from 1.5 kg of ground up corn cobs.[11] Treatment of furfural with aqueous NaOH disproportionates[12] the aldehyde into equal amounts of the acid **60** and the alcohol **61**. No doubt sodium borohydride would do the job better. Catalytic reduction gives the saturated alcohol **62** in 85% yield.[13]

We seem not to be getting closer to **58**, but **58** would actually result by dehydration of **62**. In real life, the alcohol **62** is turned into the chloride and a double elimination with NaNH$_2$ gives **58** after acidification.[14] Though we saw elimination reactions used to make alkenes in chapter 15, this is the first we have seen to make alkynes.

Now the three-membered ring can be closed, after replacing the OH group by Cl, by treatment with two equivalents of butyl-lithium. The first proton is removed from the alkyne so the cyclisation occurs on the dilithium derivative[15] **64**.

If the cyclisation is not worked up, the product is the alkynyl-lithium that can be added directly to **55** to give the alcohol **54**. In the Merck synthesis of efavirenz, this step is used to make a single enantiomer of **64** and this chemistry is discussed in *Strategy and Control*.[16]

Alkynes give us new strategies for making alkenes and ketones with disconnections in different places from those used in chapters 13 and 15.

References

1. G. H. Whitfield, *Brit. Pat.*, 1955, 735,188;   *Chem. Abstr.*, 1956, **50**, 8721f.

2. H. Pasedach, *Ger. Offen.*, 1972, 2,047,446;   *Chem. Abstr.*, 1972, **77**, 4876.

3. A. W. Johnson, *J. Chem. Soc.*, 1946, 1014.

4. J. W. Copenhaver and M. H. Bigelow, *Acetylene and Carbon Monoxide Chemistry*, Reinhold, New York, 1949, pages 130–142.

5. G. Büchi and B. Egger, *J. Org. Chem.*, 1971, **36**, 2021.

6. C. A. Henrick, *Tetrahedron*, 1977, **33**, 1845;   B. A. Bierl, M. Beroza and C. W. Collier, *Science*, 1970, **170**, 87.

7. A. A. Kraevskii, I. K. Sarycheva and N. A. Preobrazhenskii, *Zh. Obsch. Khim.*, 1963, **33**, 1831; *Chem. Abstr.*, 1964, **61**, 14518f.

8. M. S. Newman and W. R. Reichle, *Org. Synth. Coll.*, 1973, **5**, 1024.

9. G. Stork and R. Borch, *J. Am. Chem. Soc.*, 1964, **86**, 935; 936.

10. A. Thompson, E. G. Corley, M. F. Huntington, E. J. J. Grabowski, J. F. Remenar and D. B. Collum, *J. Am. Chem. Soc.*, 1998, **120**, 2028.

11. R. Adams and A. Vorkee, *Org. Synth. Coll.*, 1941, **1**, 280.

12. W. C. Wilson, *Org. Synth. Coll.*, 1941, **1**, 276.

13. H. E. Burdick and H. Adkins, *J. Am. Chem. Soc.*, 1934, **56**, 438.

14. E. R. H. Jones, G. Eglinton and M. C. Whiting, *J. Chem. Soc.*, 1952, 2873; *Org. Synth. Coll.*, 1963, **4**, 755.

15. E. G. Corley, A. S. Thompson and M. Huntington, *Org. Synth.*, 2000, **77**, 231.

16. *Strategy and Control*, pages 515, 591.

17 Two-Group C–C Disconnections I: Diels-Alder Reactions

Background Needed for this Chapter Reference to Clayden, *Organic Chemistry:* Chapter 35: Pericyclic Reactions I: Cycloadditions

The Diels-Alder reaction,[1] e.g. **1** + **2**, is one of the most important reactions in organic synthesis because it makes two C–C bonds in one step and because it is regio- and stereoselective. It is a pericyclic reaction between a conjugated diene **1** and an alkene **2** or **4** (the dienophile) conjugated with, usually, an electron-withdrawing group Z forming a cyclohexene **3** or **5**.

The disconnection is often best found by drawing the reverse reaction mechanism. You may draw the arrows either clockwise or anticlockwise but one must start from the alkene. It makes sense to draw this arrow first. The disconnection is **5a** for the general case and **6** for a specific case, revealing a diene **7** and a dienophile **8**. These reagents **7** and **8** need only to be heated together in a sealed tube[2] (because they are volatile) to give **6**.

This is a two-group disconnection because it can be carried out only when two features are present in the target molecule: the cyclohexene ring and the electron-withdrawing group outside the ring and on the opposite side to the alkene. The *relationship* between these features must be recognised. No matter how complicated a molecule may be, if those features are present we should first try a Diels-Alder reaction. Other features, such as the two four-membered rings in **9**, shouldn't distract you. In fact we made diene **10** in chapter 15 and the dienophile **11** will be discussed later. Combining the two does indeed give **9** which was used to make[3] the highly strained benzene **12**.

Organic Synthesis: The Disconnection Approach. Second Edition   Stuart Warren and Paul Wyatt
© 2008 John Wiley & Sons, Ltd

Stereospecificity

The reaction occurs in one step so there is no chance for either the diene or the dienophile to rotate and the stereochemistry of each must be faithfully reproduced in the product.[4] The two Hs in **3** are *cis* because they were *cis* in the starting anhydride **2**. The two Hs in **14** are *trans* because they were *trans* in the diester **13**.

The synthetic attractant siglure **15** used as bait for the Mediterranean fruit fly has all the features of a Diels-Alder adduct and we need the *E*-unsaturated ester **16** for the reaction.[5]

In manufacture it is easier to use the cheap methyl ester and exchange the esterifying group after the Diels-Alder reaction.

Stereospecificity of the Diene

The stereochemistry of the diene is also faithfully reproduced in the product but this is not so easy to see. Diene **19** with two *E* double bonds, adds to the acetylene **20** to give a product **21** with the two phenyl groups *cis*. This is because the two reagents approach in parallel planes **22**. There are two other ways to work this out, both based on a diagram **23** that looks downwards onto the planes. You might see straightaway that the two marked Hs are *cis* and therefore the two Ph groups must also be *cis*. You might prefer to see that both reagents are symmetrical, having a plane of symmetry marked with the dashed line in **23** and that the product must keep the same symmetry. A more detailed analysis of all these questions appears in Ian Fleming's books.[6]

19   20   21   22   23

These two aspects of the Diels-Alder are stereo*specific* in that the stereochemistry of the product is determined only by the stereochemistry of the reagents and not at all by how favourable one or other pathway may be. We deliberately used an acetylenic dienophile in the last example as it gives rise to no new stereochemistry. But if *both* diene *and* dienophile lead to new stereochemistry in the product, we need to consider stereo*selectivity* as there will be two ways in which the integrity of both reagents can be maintained.

Endo-Selectivity

In a classic Diels-Alder reaction, cyclopentadiene **24** combines with maleic anhydride **2** with complete stereospecificity to give either **25** or **26**: so the two Hs on **2** remain *cis* to each other in both **25** and **26**. These are called *endo*- and *exo*- adducts. This refers to the relationship between the alkene on the diene side and the carbonyl groups from the dienophile. These are much closer in the *endo*-adduct **25**. The result is easy to see when both reagents are cyclic.

24   2   25 (*endo*-)   or   26 (*exo*-)

The experimental result is that the *endo*-adduct **25** is kinetically favoured while the *exo*-adduct is more stable. This suggests an attractive interaction between the carbonyl groups and the middle of the diene. Indeed part of the role of the electron-withdrawing groups in the dienophile is to attract the diene through space. This interaction does not lead to any bonding between these atoms and is a secondary orbital interaction. You may find this easier to see in the 3D diagrams **27** and **28** where dotted lines show the secondary orbital interactions or in the flat diagram **29**.

endo   exo   secondary orbital-interactions   bonding inter-actions

27 gives 25   28 gives 26   29

In open chain examples, diagrams in the style of **23** or **29** are usually easier to follow, but the choice is yours. The product from **30** and acrolein is clearly **31: top tip number 1**—always draw the mechanism first. But what is the stereochemistry at the circled centres? Putting the diene on top **32** and—**top tip number 2**—drawing in the hydrogen atoms at the new centres—you should be able to see that the three marked Hs are all on the same side. This is the RHS as drawn in **32** but unfolding (or flattening) the ring makes this the top surface **31a** and hence the other groups (two methyls and an aldehyde) are on the bottom **31b**. Here is another virtue of the Diels-Alder reaction: it generally makes the less stable diastereoisomer.

| **30** | **acrolein** | **31; stereo?** | **32** | | **31a** | **31b** |

Disconnection will be needed to discover which geometrical isomer of diene or dienophile is needed for a given product. The imide **33**, needed for Weinreb's cytochalasin synthesis[7] is easily disconnected to the imide **34** and the diene **35**. We clearly need either *E, E*-or the *Z,Z*-diene to get the two substituents on the same side. But which? All the Hs are on the same side so we need the molecules shown in diagram **36** and hence *E,E*-**35**. The synthesis of this diene is discussed in the workbook for chapter 15.

| **33** | **34** | **35** | **36** | **E,E-35** |

Regioselectivity

So far we have used at least one symmetrical component but when both components in a Diels-Alder reaction are unsymmetrical, regioselectivity is an issue. A full explanation is beyond the scope of this book and you are referred to Ian Fleming's books and *Clayden* chapter 25. We need a quick way to work out what happens between a given diene and dienophile and the simplest mnemonic is that the Diels-Alder reaction has an aromatic transition state (true: six delocalised π-electrons) and that it is '*ortho,para*' directing. So in the first reaction, with a 1-substituted butadiene **37**, we get the '*ortho*' product **39** while in the second reaction, with a 2-substituted butadiene **41**, we get the '*para*' product **42**. Neither reaction gives the '*meta*' product **40**. Notice that these reactions are catalysed by a Lewis acid $SnCl_4$. This complexes to the oxygen of the ketone making the enone more polarised and enhancing the regioselectivity.

A slightly more rational way to say the same thing is that we do really know which component supplies the HOMO ('nucleophile') and which the LUMO ('electrophile'). The enone **38** is naturally electrophilic as in **43** and **45**, especially when bound to the Lewis acid. If the diene **37** acted as a nucleophile, it would give the more highly substituted allylic cations **44** and **46**. The Diels-Alder is *not* an ionic reaction and **44** and **46** are *not* intermediates but the HOMO and LUMO that determine the regiochemistry in the imaginary ionic reactions **43** and **45** also determine the regiochemistry of the pericyclic reactions.

Stereospecificity, stereoselectivity and regioselectivity combined in Diels-Alder reactions give unprecedented control and you should now see why it is so important. The analgesic tilidine **47**, effective in cases of severe pain, is an obvious Diels-Alder product.[8] The regioselectivity is correctly '*ortho*' and the *endo* transition state **51** shows that the *trans*-enamine **49** is needed. This is the geometry we get when the enamine is made in the normal way from the enal **50** and Me_2NH.

FGI on Diels-Alder Products

The cyclic ether **52** comes from the diol **53** that can be made by reduction of various Diels-Alder adducts such as the anhydride **54**.

Be careful not to attempt a synthesis based on the direct disconnection **52a** as the unsaturated ether **55** lacks the vital carbonyl group and does not react with **41**. However maleic anhydride does react, LiAlH$_4$ reduction gives the diol **53** and the cyclisation occurs[9] on treatment with TsCl and NaOH. No doubt the mono-tosylate is formed and rapidly cyclises.

Intramolecular Diels-Alder Reactions

As is usually the case, intramolecular reactions are easier than intermolecular and often do not obey the usual rules. Some do not need the carbonyl group, some show *exo* rather than *endo* selectivity, and the cyclisation of **56** gives the '*meta*' product **58**. The mechanism **57** makes it clear that the expected '*para*' product (cf. **42**) cannot be formed. This is a particularly impressive example as the product **58** is a 'bridgehead' alkene with a strained geometry.[10] The alkene is *cis* inside the six-membered ring but *trans* in the outer 10-membered ring.

Diels-Alder Reactions in Water

In an ideal world all chemical reactions would be carried out in water because solvent is the main by-product of all chemical processes and is difficult to recycle. For many reactions water as solvent is virtually impossible as reagents and/or catalysts are incompatible and/or insoluble in water. But Diels-Alder reactions are faster and more stereoselective in water even though the reagents generally don't dissolve.[11] So cyclopentadiene **24** adds to methyl acrylate **59** with poor *endo* selectivity in cyclopentadiene as solvent. The selectivity improves in ethanol but is excellent in water.[12] One explanation is that the reagents cluster in small oily droplets in water and are held closer together than they would be if they were in solution.

	endo:exo ratio in solvents	
in 24:	3.9:1	
in EtOH:	8.5:1	
in water:	21.4:1	

References

1. M. C. Kloetzel, *Org. React.*, 1948, **4**, 1; H. L. Holmes, *Ibid.*, 60; L. W. Butz, *Ibid.*, 136; J. Sauer, *Angew. Chem., Int. Ed. Engl*, 1966, **5**, 211.
2. O. Diels and K. Alder, *Liebig's Ann. Chem.*, 1929, **470**, 62.
3. R. P. Thummel, *J. Am. Chem. Soc.*, 1976, **98**, 628; J. G. Martin and R. K. Hill, *Chem. Rev.*, 1961, **61**, 537.
4. F. V. Brutcher and D. D. Rosenfeld, *J. Org. Chem.*, 1964, **29**, 3154.
5. N. Green, M. Beroza and S. A. Hall, *Adv. Pest Control Res.*, 1960, **3**, 129.
6. Ian Fleming, *Orbitals*; Ian Fleming, *Pericyclic Reactions*, Oxford University Press, 1999.
7. M. Y. Kim and S. M. Weinreb, *Tetrahedron Lett.*, 1979, 579.
8. G. Satzinger, *Liebig's Ann. Chen.*, 1969, **728**, 64.
9. N. L. Wendler and H. L. Slates, *J. Am. Chem. Soc.*, 1958, **80**, 3937.
10. K. J. Shea and P. D. Davis, *Angew. Chem. Int. Ed. Engl.*, 1983, **22**, 419.
11. H. C. Hailes, *Org. Process Res. Dev.*, 2007, **11**, 115.
12. R. Breslow, U. Maitra and D. Rideout, *Tetrahedron Lett.*, 1983, **24**, 1901.

18 Strategy VIII: Introduction to Carbonyl Condensations

The next 10 chapters are about the synthesis of carbon skeletons with two functional groups. Compounds such as **1–3** will all be treated as 1,3-difunctionalised compounds since the important thing is not the type of functional group but the relationship between them. Our logic is that all FGs can be derived from alcohols, ketones (or aldehydes) or acids by substitution and that those three can be interconverted by oxidation or reduction.

Analysis will be by FGI to reveal the oxygen-based functionality at the right oxidation level and then by C–C disconnection using the relationship between the FGs as our guide. So we shall be using two-group disconnections throughout. The carbon synthons used will be the same as those we used for two-group C–X disconnections in chapter 6. Most of the chemistry revolves around the carbonyl group and we need to think first how that group affects the behaviour of molecules.

The carbonyl group is the most important functional group in organic synthesis because it can be naturally electrophilic or nucleophilic at carbon. It hardly needs saying that carbonyl compounds are naturally electrophilic, **4** or **5**, at the carbonyl carbon and so react with nucleophiles at that atom **6**. If there is a leaving group X, the tetrahedral intermediate pushes it out **7** to regenerate the carbonyl group. The result **8** is acylation of the carbon nucleophile.

The corresponding disconnection is of the newly formed C–C bond **8a**. The synthons are the acyl cation and a nucleophilic carbon species that might be a metal derivative RM (chapter 13) but will generally be an enolate in the next 10 chapters. And that is how carbonyl compounds are nucleophilic.

Organic Synthesis: The Disconnection Approach. Second Edition   Stuart Warren and Paul Wyatt
© 2008 John Wiley & Sons, Ltd

Carbonyl compounds such as acetone **10** exist predominantly in the keto form **10** but are in equilibrium with the enol form **11**. We shall be more interested in the formation of the enolate anion **13** with base **12** and its reactions at the α-carbon with carbon electrophiles.

The disconnection is of the newly formed C–C bond **14a** and is not the same as **8a**. The synthons are represented by the enolate anion and a carbon electrophile. We saw alkyl halides in this role in chapter 13 but in the next 10 chapters we shall be mostly interested in combining enol(ate)s with carbonyl compounds.

Nucleophilic and Electrophilic Synthons

The synthons **9** and **15** have natural polarity and it will be helpful in the next 10 chapters if you recognise whether synthons such as these and **17** have natural polarity. We have discussed **9** and **15** above and **17** in chapters 6 and 13 where we used α,β-unsaturated compounds **18** as electrophiles. You may find it helpful to use the labels suggested by Seebach.[1] The letters **a** and **d** are used for acceptor (electrophilic) and donor (nucleophilic) synthons and a superscript number shows which atom is meant. The carbonyl group is always atom number 1. So the enolate becomes a d^2 synthon and **9** and **17** are a^1 and a^3 synthons. If you don't like these labels, ignore them, they are entirely optional.

So if we wish to make a compound like **19** we can disconnect a C–C bond to reveal two synthons, **20**, a d^2 synthon easily recognised as the enolate of ketone **22**, and an a^1 synthon **21** realised in the aldehyde **23**. We have rediscovered the aldol reaction.

If we wish to make the diketone **24**, disconnection to the same enolate **20** reveals the **a³** synthon **25** and we already know that enone **26** is the reagent. Both these syntheses use synthons of natural polarity: the enolate anion of one compound **22** and either a simple carbonyl compound **23** or the conjugated enone **26**.

For this reason, our chapters on two group C–C disconnections follow a slightly odd order. First we deal with the odd numbered relationships: the 1,3-diCO **19a** (chapter 19) and the 1,5-diCO **24a** (chapter 21) and then we turn to the even-numbered relationships 1,2-diCO **27** (chapter 23) and 1,4-diCO **28** (chapter 25) because these will need synthons of unnatural polarity. Finally we shall turn to the 1,6-diCO relationship (chapter 27) as that involves a totally different strategy.

We can summarise this in these three points:

1. Synthesis of difunctionalised compounds with *odd* numbered relationships needs only synthons of *natural* polarity.
2. Synthesis of difunctionalised compounds with *even* numbered relationships needs some synthons of *unnatural* polarity.
3. All odd numbered acceptor synthons (such as **a¹** and **a³**) and even numbered donor synthons (such as **d²** and **d⁴**) have unnatural polarity.

And add a slogan: **Before you do C–C disconnections, COUNT the relationships**.

You will notice that all these methods depend on the carbonyl group and interspersed with these chapters will be strategy chapters relevant to them culminating in a general strategy chapter on carbonyl chemistry:

Chapter 20: Strategy IX: Control in Carbonyl Condensations.
Chapter 22: Strategy X: Use of Aliphatic Nitro Compounds in Synthesis.
Chapter 24: Strategy XI: Radical Reactions in Synthesis: FGA and its Reverse.
Chapter 26: Strategy XII: Reconnections.
Chapter 28: General Strategy B: Strategy of Carbonyl Disconnections.

Carbon Acids and the Bases Used to Deprotonate Them

We shall be using a variety of bases to create the enolate anions used in the next few chapters and it helps if you have some idea of relative base strength. In the table any base can be used to deprotonate a carbon acid lower in the table, that is, the conjugate acid of the base should have a higher pK_a than the carbon acid. Do not try to learn all these numbers but a general idea of the magnitudes will help you. You could refer to Clayden, *Organic Chemistry*, chapter 8 for the basics. Among the carbon acids, the proton(s) in italics are the ones removed by a base. You should realise that pK_a outside the range of water (pH about 0–15) are determined indirectly and there may not be in agreement about the exact number.

TABLE 18.1 Carbon acids and the bases used to deprotonate them

Carbon Acid	pK_a	Base B	pK_a of BH	From
Alk-*H*	~42	BuLi	42	available
		RMgBr		RBr + Mg
Ar-*H*	~40	ArLi	~40	PhLi available
CH$_2$=CHC*H*$_3$	38			
PhC*H*$_3$	37	NaH	~37	available
		i-Pr$_2$NLi (LDA)		*i*-Pr$_2$NH + BuLi
MeSO.C*H*$_3$ (DMSO)	35	MeSO.CH$_2^{\ominus}$ $^{\ominus}$NH$_2$	35	DMSO + NaH Na + NH$_3$(l)
Ph$_3$C*H*	30	Ph$_3$C$^-$	30	
HC+C*H*	25			
C*H*$_3$CN	25			
C*H*$_3$CO$_2$Et	25			
C*H*$_3$COR	20			
C*H*$_3$COAr	19	*t*-BuOK	19	available
Ph$_3$P$^+$-C*H*$_3$	18	EtO$^{\ominus}$, MeO$^{\ominus}$	18	ROH + Na(0)
ClC*H*$_2$COR	17			
PhC*H*$_2$COPh	16	HO$^{\ominus}$	16	available
MeCOC*H*$_2$CO$_2$Et	11			
C*H*$_3$NO$_2$	10	PhO$^{\ominus}$	10	PhOH + NaOH
		Na$_2$CO$_3$		available
		amines: R$_3$N etc		
EtO$_2$CCH$_2$CN	9			
Ph$_3$P$^+$ C*H*$_2$CO$_2$Et	6	NaHCO$_3$	6	available
		AcO$^{\ominus}$		available
		pyridine		available

Reference

1. D. Seebach, *Angew. Chem. Int. Ed. Engl.*, 1979, **18**, 239.

19 Two-Group C–C Disconnections II: 1,3-Difunctionalised Compounds

Background Needed for this Chapter References to Clayden, *Organic Chemistry:* Chapter 27: The Aldol Reaction; Chapter 28: Acylation at Carbon.

This chapter deals with target molecules of two main types: hydroxyketones **1** and 1,3- or β-diketones **4**. Both have a 1,3-relationship between the two functionalised carbons. Both can be disconnected at one of the C–C bonds between the functional groups to reveal the enolate **2** of one carbonyl compound reacting with either an aldehyde **3** or acid derivative **5** such as an ester.

We shall need to understand the formation of enol(ate)s from aldehydes, ketones and esters and it is worthwhile establishing now that these three types of compounds form a graded series of electrophiles whilst their enolates form a graded series of nucleophiles in the reverse direction. Any of these enolates can react with any of the carbonyl compounds.

β-Hydroxy Carbonyl Compounds: The Aldol Reaction

With compounds of type **1**, only one of the two C–C bonds is worth disconnecting: the one next to the hydroxyl carbon. A simple example without any selectivity is ketone **6** which disconnects to the enolate **7** and the ketone **8**. It is easy to see that **7** is the enolate of **8** so this is a 'self-condensation': we simply need to create a small amount of enolate **7** in the presence of much unenolised ketone **8** and the reaction will occur.

Bases like hydroxides or alkoxides are of about the right strength: barium hydroxide is often used.[1] The small amounts of enolate **7** quickly add to the large excess of ketone **8** to give the anion **9** of the product which regenerates the base by taking a proton from water or the alcohol.[2]

Products such as **6** are 'aldols' having OH and CHO groups and this reaction is the aldol reaction. The diol **11** that is used to make Meyers' heterocycle[3] **10** is not an aldol but FGI of the only alcohol that could come from a ketone, reveals an aldol product: in fact the dimer of acetone.

The aldol reaction uses barium hydroxide and the reduction that could be carried out with many reducing agents, works well catalytically.[4]

Wanting to study the photochemistry of a β,γ-unsaturated ketone that could not isomerise to a conjugated ketone, chemists[5] chose **13**. The obvious Wittig disconnection revealed the ketoaldehyde **14**. Observing the symmetry of the carbon skeleton and the 1,3-relationship, they changed the ketone into an alcohol so that an aldol disconnection revealed two molecules of aldehyde **16**.

A small amount of sodium hydroxide was enough for the aldol, and oxidation by CrO_3 in pyridine was chemoselective. You might have supposed that the Wittig with a stabilised ylid would give the *E*-alkene **13** selectively but in fact it gave a 50:50 mixture of *E*-**13** and *Z*-**13**. This was no problem as they could be separated and the chemists wanted to study the photochemistry of both.

The Synthesis of α,β-Unsaturated Carbonyl Compounds

The impossibility of a conjugated alkene in **15** as well as **13**, makes this an exceptional case in aldols of aldehydes. Without the branch at the α-carbon, the product is more usually a conjugated enal.[6] So the linear isomer **17** of **16** with the same base gives the enal **18** in good yield.[7] The true product of the dimerisation is the anion **19**. This is in equilibrium with the enolate that allows an E1cB elimination of water **20** to give the enal **18**.

The first disconnection for any α,β-unsaturated carbonyl compound **21** is an FGI reversing the dehydration. We could suggest two alcohols: **22** or **25** but we much prefer the 1,3-diO relationship in **22** to the 1,2-diO in **25** as the synthesis of compounds with odd numbered relationships needs synthons of only natural polarity (chapter 18).

This sequence is a classical 'condensation'—two molecules react together with the extrusion of a small molecule (water in this case) but the term is now used of most carbonyl reactions of this sort. Another example is that of lactones such as **26**. Disconnection reveals two molecules of the simple lactone **28** and condensation with NaOMe in MeOH gives **26** in >66% yield.[8]

Having got the idea, you might not want to be bothered with the rehydration step as it is easy to see the hidden carbonyl group where the alkene is and the half of the molecule with the carbonyl group must be the enolate in real life. Most people just disconnect the alkene and write the two starting materials in one step. So the lactone disconnection becomes **26a** and the general case **21a**. This again is a matter of personal choice.

Intramolecular Aldol Reactions

Perfect selectivity is the result when symmetrical dialdehydes or diketones cyclise with five- and six-membered rings being preferred to the rest. The linear diketone **29** cyclises to the cyclohexene **30**: it doesn't matter which of the identical α-positions enolises as it will always attack the other ketone.[9] Cyclisation of cyclodecadione **31** is even more impressive: there are four equivalent enolisable α-positions but all give[10] the same product **32**.

1,3-diCarbonyl Compounds

When we come to 1,3-dicarbonyl compounds **4** the principle is the same but we now have a choice: the keto-ester **35** could be disconnected **35b** to the enolate **36** of acetone and diethyl carbonate **37** and this synthesis would work but we prefer **35a** as that gives us the enolate of ethyl acetate **34** and ethyl acetate itself **33**—another self condensation.

There is now a reason for choosing one particular base: ethoxide ion. It will produce a small amount of the enolate **34** that will react with unenolised ester **40** to give the product **35**, regenerating the base. That is useful, but the real reason we use the same base as the esterifying alcohol (i.e. ethoxide with an ethyl ester) is that if it attacks as a nucleophile **38** instead of as a base **39** it merely regenerates ethyl acetate. Reacting ethyl acetate with sodium ethoxide in ethanol gives ethyl acetoacetate **32** in 60% yield.[11] This ketoester is therefore available cheaply and is used as a starting material in many syntheses (chapter 13).

Intramolecular reactions are very favourable here as elsewhere and the cyclic ketoester **41** can be disconnected to the symmetrical diester **42**. The cyclisation works well.[12] In these reactions the proton between the two carbonyl groups of the product (marked in **41**) is removed by the ethoxide formed in the reaction to give the stable enolate **43**. This is easily demonstrated by adding an alkyl halide in the work-up when products **44** are formed.

Since β-ketoesters like **41** and **44** can be decarboxylated easily, it makes sense to use this efficient cyclisation wherever we can. You might first think of disconnecting cyclic ketone **46** to MeNH$_2$ and divinyl ketone **45** but this looks a rather unstable compound. If we add a CO$_2$Et group, we can use our 1,3-diCO disconnection to symmetrical **48** and only then revert to the 1,3-diX disconnection as both starting materials **46** are available.

The synthesis is just that,[12] though NaH was used for the cyclisation and the decarboxylation accomplished by hydrolysis of the ester and heating with 20% HCl.

Looking Forward

Most of the examples in this chapter have been of molecules without selectivity. They have indeed all been self condensations. We hope this has established the basic disconnections and the chemistry but we must now turn to examples where selectivity is needed. So the ketone **46** was made to study aldol reactions with aromatic aldehydes.[13] They found that, in acid or base, the enone **52** was the main product with the best yield from HCl in EtOH. The product **52** was isolated as its HCl salt. In this case it is easy to see that only the ketone can enolise, that the aldehyde is more electrophilic than the ketone and that the geometrical isomer shown is the more stable. Such considerations are the substance of the next chapter.

References

1. *Vogel*, page 798.
2. A. T. Nielsen and W. J. Houlihan, *Org. React.*, 1968, **16**, 1; see page 115.
3. A. I. Meyers, *Heterocycles in Organic Synthesis*, Wiley, 1974.
4. J. B. Conant and N. Tuttle, *Org. Synth. Coll.*, 1932, **1**, 199; H. Adkins and H. I. Cramer, *J. Am. Chem. Soc.*, 1930, **52**, 4349.
5. W. G. Dauben, M. S. Kellogg, J. I. Seeman and W. A. Spitzer, *J. Am. Chem. Soc.*, 1970, **92**, 1786.
6. A. T. Nielsen and W. J. Houlihan, *Org. React.*, 1968, **16**, 1, table II, pages 86–93.
7. *Vogel*, page 802;   Clayden, *Organic Chemistry*, chapter 19.
8. O. E. Curtis, J. M. Sandri, R. E. Crocker and H. Hart, *Org. Synth. Coll.*, 1963, **4**, 278; H. Hart and O. E. Curtis, *J. Am. Chem. Soc.*, 1956, **78**, 112.
9. E. E. Blaise, *Bull. Chim. Soc. Fr.*, 1910, **7**, 655.
10. Ref 2, table VI, page 125.
11. J. H. Inglis and K. C. Roberts, *Org. Synth. Coll.*, 1932, **1**, 235.
12. P. S. Pinkney, *Org. Synth. Coll.*, 1943, **2**, 116; W. Dieckmann, *Ber.*, 1894, **27**, 102; J. P. Schaefer and J. J. Blomfield, *Org. React.*, 1967, **15**, 1.
13. S. M. McElvain and K. Rorig, *J. Am. Chem. Soc.*, 1948, **70**, 1820.

20 Strategy IX: Control in Carbonyl Condensations

Background Needed for this Chapter References to Clayden, *Organic Chemistry:* Chapter 27: The Aldol Reaction; Chapter 28: Acylation at Carbon.

The last chapter introduced some good disconnections based on carbonyl compounds as both nucleophiles and electrophiles but avoided all questions of chemo- or regioselectivity. These reactions are so important that you need to understand how to control these issues. All the chief difficulties crop up in the synthesis of the conjugated enone **1**.

This all looks very sound and we met examples in the last chapter. We want the ketone **2** to form an enolate and to combine **4** with the aldehyde **3** to give the anion of an aldol that would almost certainly dehydrate to the target molecule **1**.

But will any of this happen? We want the ketone **2** to form the enolate, but won't the aldehyde **3** form an enolate more easily? We want the enolate to form on the less substituted side of the ketone, but won't the conjugated enolate be more stable? We want the enolate to attack the aldehyde, but might it not instead attack another molecule of **2** in a self-condensation? We want a cross-condensation between two different carbonyl compounds. To be satisfied with our plan we need answers to these three questions:

The Three Key Questions for Successful Cross-Condensations

1. Which compound will enolise/form an enolate?
2. If it is an unsymmetrical ketone: on which side will it form an enol(ate)?
3. Which compound will act as the electrophile?

Organic Synthesis: The Disconnection Approach. Second Edition   Stuart Warren and Paul Wyatt
© 2008 John Wiley & Sons, Ltd

Fortunately it is rare that all three questions are important, but, even if they are, methods are now available to deal with all but the most intractable cases. In general the problems arise because of the relative reactivity of the main classes of carbonyl compound. This diagram is not quite the same as the diagram in the last chapter both because we need to add all the acid derivatives and also because we need to emphasise something slightly different.

Since the same compounds are most easily enolised and most electrophilic, they tend to self-condense rather than react with anything else. So that, in the reaction of **2** and **3** under equlibrating conditions, the aldehyde **3** will probably self condense and ignore the less reactive ketone. This chapter looks at ways to overcome that tendency. We examined self-condensation in chapter 19 so we shall look at other cases now.

Intramolecular Reactions

These are the easiest to control as five- and six-membered rings are preferred. If that is what we want, we should use the equilibrium methods we have met so far and allow the molecule to find its own way to the most stable product. Four different enolates **7, 8, 12** and **13** could be made from the diketone **10** by removal of four different protons. Each could cyclise onto the other carbonyl group to give three-membered rings **9** or **11** or five-membered rings **6** or **14**. These alkoxides are all in equilibrium with **10** via the enolates and so the unstable three-membered rings quickly revert to **10**. So which compound is formed: **6** or **14**?

Only product **16** from intermediate **14** is formed.[1] This is partly because **14** contains two fused virtually strain-free five-membered rings while **6** has strained bridged rings (these are by no means impossible: we saw them in chapter 17) but mainly because the E1cB elimination we saw in chapter 19 can occur only on **14**. Elimination of water from **6** would give an impossible bridgehead alkene. We can draw the full mechanism more realistically as **13a** to **16**.

So to summarise: cyclisations to strained three- and four-membered rings are normally reversible and stable five- or six-membered rings are usually preferred, especially if elimination of water makes a stable conjugated product.

Cross-Conjugations I: Compounds that Cannot Enolise

If a compound cannot enolise because it has no protons on the α-carbon atoms then it can take part in a carbonyl condensation only as the electrophile. This will be useful only if it is strongly electrophilic (to avoid self-condensation of the other compound, see point 3 above) so you will see mainly aldehydes and acid chlorides in this list.

Carbonates are useful for adding the CO_2Et group to make stable enolates of the kind we met in chapter 13 and will meet again soon. Here is a case where disconnecting one carbon atom reveals an available starting material **24**. Only the ketone **24** can form an enolate and the carbonate is more electrophilic than the ketone. The ideal base is ethoxide ion to avoid ester exchange, but stronger bases such as NaH work well too.[2]

This method is useful for aryl-substituted malonates **27** where the normal alkylation of the malonate anion **26** is impossible as S_N2 reactions fail on unactivated aryl halides. For Ar = Ph reaction of **28** with NaH and diethyl carbonate gives[3] **27**; Ar = Ph in 86% yield.

Aromatic aldehydes condense well with aliphatic ketones such as acetone to give either the mono- **31** or di-adducts **29** depending on conditions.[4] In an excess of acetone, **31** is the main product but in ethanol with two equivalents of benzaldehyde, the diadduct **29** is the only product. This **29** is dibenzylidene acetone or dba, an important ligand for palladium.

Further selectivity is needed if the enol component is an unsymmetrical ketone. Some selectivity can be achieved by choice of acid, favouring the more substituted enol, or base, favouring kinetic enolate formation on the less substituted side. The acid **32** was used at a very early stage of Woodward and Eschenmoser's synthesis[5] of vitamin B$_{12}$. Standard α,β-unsaturated carbonyl disconnection revealed unsymmetrical ketone **33** and unenolisable but very electrophilic glyoxylic acid **34** available as its hydrate. In acid solution reaction occurred very selectively indeed.

By contrast, Woodward's synthesis[6] of the supposed structure **36** of the antibiotic patulin, which later turned out to have the isomeric structure **35**, required intermediate **37** drawn both to resemble **36** and in a more normal way. Woodward then made the correct structure **35**.

Disconnection of the 1,3-diCO relationship **37a** gave the unsymmetrical ketone **38** and the unenolisable, symmetrical, and very electrophilic (again two carbonyl groups joined together: very electrophilic) oxalate **23**; R = Me. Now enolate formation needs to occur on the methyl group rather than the more substituted side. The answer was to use base.

This is thermodynamic control. The initial products **37** and **41** are converted into the stable enolates **40** and **42** by the methoxide released in the reaction. The enolate **40** is more stable than **42** as it has one fewer substituents.

Formaldehyde: the Mannich Reaction

One obvious candidate for an electrophilic but non-enolisable compound is formaldehyde $CH_2=O$ but it is simply too electrophilic to be well controlled. A trivial example is its reaction with acetaldehyde and hydroxide ion. The first aldol gives the expected product **43** but a second gives **44** and a third follows. Now hydroxide adds to another molecule of formaldehyde and delivers a hydride ion **45** in the Cannizzaro reaction (the other product is formate ion HCO_2^-) to give 'pentaerythritol' **46**, a useful compound in polymer chemistry for cross-linking but not much use to us. We need to moderate the unruly behaviour of this useful one-carbon electrophile.

We need a formaldehyde equivalent that is less electrophilic than formaldehyde itself and will therefore add only once to enol(ate)s. The solution is the Mannich reaction.[7] Formaldehyde is combined with a secondary amine to give an iminium salt that adds **47** to the enol of the aldehyde or ketone in slightly acidic conditions to give the amino ketone (or 'Mannich base') **48**. If the product of the aldol reaction **50** is wanted, alkylation on nitrogen provides a good leaving group and E1cB elimination does the trick.

Sometimes the hydrochloride of the Mannich base eliminates on heating and the vinyl ketone **53** can be made this way.[8]

If base is needed, it may be that even sodium bicarbonate $NaHCO_3$ is strong enough as was the case in Whiting's synthesis[9] of the acetal **54**. Sequential 1,1- and 1,2-diX disconnections take us back to the enone **56**, and obvious Mannich product

The Mannich reaction gave a base that was methylated without isolation to give the salt **58**. Elimination with $NaHCO_3$ gave the enone, nucleophilic epoxidation with HO_2^- gave the epoxide

59 that was hydrolysed to the diol **55**. Finally, the acetal was made with acetone in acid solution in 72% yield.

Cross-Conjugation II: Specific Enolates

We need now to look at situations where both compounds might enolise and see how specific enolates can be used to control which compound does so (chemoselectivity) before looking at how we control which side of an unsymmetrical ketone forms the enolate (regioselectivity). We met two specific enol equivalents in chapter 13: β-dicarbonyl compounds and lithium enolates and they are the keys to this section.

β-*Dicarbonyl Compounds as Specific Enols in Carbonyl Condensations*

Attempts to react enol(ate)s of esters with aliphatic aldehydes are doomed as the aldehyde will simply condense with itself. If the ester is replaced by a malonate **60**, there is so much enol(ate) from the β-dicarbonyl compound that the reaction is good. This style of aldol reaction is often called a Knoevenagel reaction[10] and needs only a buffered mixture of amine and carboxylic acid. The enol reacts with the aldehyde **61** in the usual way and enolisation of the product **62** usually means that dehydration occurs under the conditions of the reaction.

If the product **63** is hydrolysed and heated in acid, decarboxylation occurs to give the unsaturated acid and re-esterification gives the ester that might have been the product from condensation of ethyl acetate and acetaldehyde.

If the original reaction is carried out under more vigorous conditions with malonic acid **67**, the decarboxylation occurs during the reaction to give the unsaturated acid in one step. This is a simple way to make[11] substituted cinnamic acids **68**.

Any combination of two carbonyl or other electron-withdrawing groups will do this reaction. Compound **69** was needed for a barbiturate synthesis. As cyanide is very anion-stabilising, disconnection gives the ketone **70**, whose synthesis will be discussed in chapter 30, and the nitrile **71**. The synthesis was straightforward once the right conditions had been found.[12]

Lithium Enolates

Lithium enolates of esters **72** can be made direct from the ester itself with the strong hindered bases LDA or LiHMDS [(Me$_3$Si)$_2$NLi] and react cleanly with even enolisable aldehydes and ketones, e.g. **74**, to give aldols **75** in high yield.[13]

One difference between this method and the malonate method is that lithium enolates add direct to the carbonyl group of enals **76** while malonates do conjugate addition. Further, malonate adducts such as **62** normally dehydrate under the reaction conditions while lithium enolates normally give the aldol product **77** without dehydration.

One of the most important contributions of lithium enolates is to stereoselectivity. Very hindered esters such as **78** form the 'trans' enolate and give high selectivity[14] in favour of the *anti* aldol **80**. These stereoselective aldols are discussed in *Strategy and Control*.

Another important contribution is to the regioselectivity of enolate formation from unsymmetrical ketones. As we established in chapter 13, ketones, particularly methyl ketones, form lithium enolates on the less substituted side. These compounds are excellent at aldol reactions even with enolisable aldehydes.[15] An application of both thermodynamic and kinetic control is in the synthesis of the gingerols, the flavouring principles of ginger, by Whiting.[16]

Gingerol-6 **81** is an obvious aldol product and disconnection reveals an unsymmetrical ketone **82** and an enolisable aldehyde. We get no favourable answer to any of the three questions at the start of this chapter: control is needed. The ketone **82** could be made in many ways but FGA to the enone **84** allows a second aldol disconnection and reveals vanillin **85** and acetone as very cheap starting materials.

The first aldol needs no control: only acetone can enolise and the aldehyde **85** is more electrophilic than acetone. Aldol reaction under equilibrating conditions gives **84** in good yield and the alkene can be reduced catalytically. The ketone **82** also contains an acidic phenol so that must be protected before the next aldol and a silyl group is the answer **86**.

The lithium enolate **87** forms almost exclusively on the methyl side – less than 4% of the other isomer could be detected – and reaction with the aldehyde **83** followed by deprotection in aqueous acid gives gingerol.

Wittig Reagents as Specific Enolates

In many ways the simplest route to an unsaturated carbonyl compound is by Wittig reaction with a stabilised ylid or a phosphonate ester (chapter 15). In his leukotriene work, Corey[17] used the stabilised ylid **90** in a reaction with the sugar deoxyribose, a hemiacetal **88** that is in equilibrium with the aldehyde **89**. Drawing the ylid as an enolate **90** makes the point that the formation of **91** is essentially an aldol condensation with Ph_3PO rather than water being lost. It is a great advantage that the ylid is so stabilised as no protection of the three OH groups is needed, unlike our last example, and the Wittig reaction may even be carried out in acid solution.

Wittig reagents can represent enolates of unsymmetrical ketones. From Corey's work on arachidonic acid metabolites[18] comes the coupling between the aldehyde **92** and the phosphonium salt **93**. This is very impressive as both components have multiple functionality and there is no loss of stereochemical integrity even though the Wittig reaction is done in aqueous NaOH.

Enamines as Specific Enols

Among the best enol equivalents for aldehydes are enamines.[19] They are stable compounds, easily made from aldehydes **95** and secondary amines, reacting with electrophiles in the same way as enols **96** to give iminium salts **97**, hydrolysed to substituted aldehydes **98**.

They are useful for ketones too. Disconnection of the enone **99** reveals an aldol reaction between cyclopentanone **74** and the enolisable ketone **100**. Control is needed solely to prevent self-condensation of the aldehyde.

The enamine **102**, made with the cyclic secondary amine morpholine **101**, does the job admirably.[20] The immediate product is the conjugated enamine **103** rather than an imine, but this is easily hydrolysed to the enone **99** with aqueous acid.

Acylation is a good way to 1,3-dicarbonyl compounds and an aldehyde example is **104** that requires a specific enol(ate) of aldehyde **105** and an acylating agent such as **106**.

This time another cyclic amine, pyrrolidine was used to make the enamine **107** and acylation occurred cleanly at carbon in spite of the formation of a quaternary centre. The wide ranging yields are for different Ar groups.[21] The intermediate is an iminium salt **108** that can be isolated. The equilibrium methods used earlier for 1,3-dicarbonyl compounds would not work here as the product **104** cannot form a stable enolate.

Cross-Condensation III: Removal of a Product from Equilibrium

We have already met this in the formation of **16** by dehydration and in the formation of **37** by stable enolate formation. A couple more examples should make the general strategy clear. The unsymmetrical ketone **110** can form an enolate on either side and at first it seems that we shall need a specific enolate to control the aldol reaction. But one product **109** cannot eliminate water while the other **111** can. Under equilibrating conditions **112** is the only product.[22]

The tricarbonyl compound **113** reacts cleanly with formaldehyde to give the lactone **115** as the first adduct **114** rapidly cyclises to the five-membered ring. The conditions are weak base and piperidine, the last of the three most popular secondary amines used in this chapter. Control is a mixture of intramolecular reaction, stable enol(ate) formation and steric hindrance.[23]

The lactone reacts regioselectively with benzaldehyde in acid solution at the methyl group rather than with the stable enol. Though **116** may be the first-formed intermediate, it cannot dehydrate while the alternative **117** can and the enone **118** is the only product. This product was needed for an alternative approach to patulin **35**.

With the range of methods for controlling carbonyl condensations now available, and we have by no means mentioned them all, it is possible to control almost any reaction. The most important message of this chapter is that you should ask yourself the three questions from the start of the chapter before plunging in to an unconsidered reaction.

References

1. H. Paul and I. Wendel, *Chem. Ber.*, 1957, **90**, 1342.

2. H. R. Snyder, L. A. Brooks and S. H. Shapiro, *Org. Synth. Coll.*, 1943, **2**, 531; A. P. Krapcho, J. Diamanti, C. Cayen and R. Bingham, *Ibid.*, 1973, **5**, 198.

3. P. A. Levene and G. M. Meyer, *Org. Synth. Coll.*, 1943, **2**, 288; G. R. Zellars and R. Levine, *J. Org. Chem.*, 1948, **13**, 160.

4. *Vogel*, page 1033.

5. A. Eschenmoser and C. E. Wintner, *Science*, 1977, **196**, 1418.

6. R. B. Woodward and G. Singh, *J. Am. Chem. Soc.*, 1949, **71**, 758; 1950, **72**, 1428; B. Puetzer, C. H. Nield and R. H. Barry, *J. Am. Chem. Soc.*, 1945, **67**, 833.

7. M. Tramontini, *Synthesis*, 1973, 303.

8. *Vogel*, page 1053.

9. A. P. Barcierta and D. A. Whiting, *J. Chem. Soc., Perkin Trans. 1*, 1978, 1257.

10. G. Jones, *Org. React.*, 1967, **15**, 204.

11. C. A. Kingsbury and G. Max, *J. Org. Chem.*, 1978, **43**, 3131; S. Rajagopalan and P. V. A. Raman, *Org. Synth. Coll.*, 1955, **3**, 425; J. Koo, G. N. Walker and J. Blake, *Ibid.*, 1963, **4**, 327, D. F. DeTar, *Ibid.*, 1963, **4**, 731.

12. J. W. Opie, J. Seifter, W. F. Bruce and G. Mueller, *U. S. Pat.*, 1951, 2,538,322; *Chem. Abstr.*, 1951, **45**, 6657c.

13. Clayden, *Lithium*; M. W. Rathke, *J. Am. Chem. Soc.*, 1970, **92**, 3223.

14. C. H. Heathcock, C. T. Buse, W. A. Kleschick, M. C. Pirrung, J. E. Sohn and J. Lampe, *J. Org. Chem.*, 1980, **45**, 1066.

15. G. Stork, G. A. Kraus and G. A. Garcia, *J. Org. Chem.*, 1974, **39**, 3459.

16. P. Deniff, I. Macleod and D. A. Whiting, *J. Chem. Soc., Perkin Trans. 1*, 1981, 82.

17. E. J. Corey, A. Marfat, G. Goto and F. Brion, *J. Am. Chem. Soc.*, 1980, **102**, 7984; E. J. Corey, A. Marfat, J. E. Munroe, K. S. Kim, P. B. Hopkins and F. Brion, *Tetrahedron Lett.*, 1981, **22**, 1077.

18. E. J. Corey, A. Marfat and B. G. Laguzza, *Tetrahedron Lett.*, 1981, **22**, 3339, E. J. Corey and W.-G. Su, *Tetrahedron Lett.*, 1984, **25**, 5119.

19. G. Stork, A. Brizzolara, H. Landesman, J. Szmuszkovicz and R. Terrell, *J. Am. Chem. Soc.*, 1963, **85**, 207.

20. L. Birkofer, S. Kim and H. D. Engels, *Chem. Ber.*, 1962, **95**, 1495.

21. S.-R. Kuhlmey, H. Adolph, K. Rieth and G. Opitz, *Liebig's Ann. Chem.*, 1979, 617; L. Nilsson, *Acta Chem. Scand. (B)*, 1979, **33**, 203.

22. A. T. Nielsen and W. J. Houlihan, *Org. React.*, 1968, **16**, 115, see page 211.

23. E. T. Borrows and B. A. Hems, *J. Chem. Soc.*, 1945, 577.

Two-Group C–C Disconnections III: 1,5-Difunctionalised Compounds Conjugate (Michael) Addition and Robinson Annelation

21

Background Needed for this Chapter Reference to Clayden, *Organic Chemistry:* Chapter 29: Conjugate Addition of Enolates.

Another odd-numbered relationship means we can still use synthons of natural polarity. The 1,5-diketone **1** disconnects to a d^2 synthon, an enolate, and an a^3 synthon **2**, that you should realise (chapter 6) is represented by the reagent **3**. The conjugation in the enone makes the terminal carbon atom electrophilic.

The only new thing in this chapter is the combination of these two reagents so that a C–C bond is made by conjugate addition of an enolate to the enone **5** giving an enolate **6** of the product that gives the 1,5-diketone **1** on protonation.

This raises the regioselectivity question of whether the enolate will add in a conjugate (or Michael) fashion **5** or directly to the carbonyl group. We need to consider which types of enol(ate) and which types of enone (Michael acceptors) are good at conjugate rather than direct addition.

The second point was made in chapter 6 where we said: 'Very electrophilic compounds such as acid chlorides or aldehydes tend to prefer direct addition while less electrophilic compounds such as esters or ketone tend to do conjugate addition.' That remains true and the same idea applies to the enolates: very nucleophilic enolates such as lithium enolates tend to prefer direct addition while less nucleophilic enols and enolates such as enamines or 1,3-dicarbonyl compounds tend to do conjugate addition.

Organic Synthesis: The Disconnection Approach. Second Edition   Stuart Warren and Paul Wyatt
© 2008 John Wiley & Sons, Ltd

Specific Enol Equivalents Good at Michael Addition

1,3-Dicarbonyl Compounds

So if we want to make **9** we have a choice between adding an enolate equivalent of the aldehyde **7** to an unsaturated ester **8** or an enolate equivalent of the ester **11** to an unsaturated aldehyde **10**. We prefer the first **9a** as the unsaturated ester **8** is more likely to do conjugate addition. An enamine would be a good choice for **7**.

However, if we make a small change in the structure by adding two methyl groups **13**, our favoured disconnection **13a** would be possible only after the discovery of five-valent carbon! We shall have to find a way to use unsaturated aldehydes as Michael acceptors.

Fortunately, we know from the last chapter how to make α,β-unsaturated carbonyl compounds so the disconnection of the enone **3** poses no problems. Both starting materials are ketones: one **4** must provide the specific enolate and the other **16** the enone **3** by the Mannich reaction.

To get conjugate addition we might use a β-ketoester **18** or an enamine for the enolate and we might carry out the reaction using the Mannich salt **19** so that the elimination will be caused by the same base that makes the enolate. Ester hydrolysis and decarboxylation of **20** would give **1**.

An example[1] that shows how good these reactions can be is the addition of the cyclopentadione **21** with acrolein in water to give the adduct **23** in 100% yield. This must be a reaction of the enol **22** and, even though the Michael acceptor is an aldehyde and a new quaternary centre is created, no acid or base is needed.

The keto-acid **24** is best disconnected at the branchpoint where the chain joins the ring giving the available cyclohexenone **25** and the enolate synthon **26** best represented by malonate **27**.

The synthesis[2] uses ethoxide as base to avoid ester exchange and the conjugate addition goes in better yield than the apparently trivial hydrolysis and decarboxylation.

Michael reactions of this sort work best when they follow a catalytic cycle. Malonate anion **28** adds to an enone to give the enolate anion **30** that collects a proton from malonate **27** and forms another molecule of the anion **28** for the next cycle.

So when Stevens[3] wanted the amino-diacetal **33** for his synthesis of coccinelline **32**, the defence compound ladybirds exude from their knees, he changed the amine into a ketone **34** with reductive amination in mind so that there would be two concealed 1,5-diCO relationships, more obvious when the acetals are removed **35**.

Since **35** is a symmetrical ketone we can use the strategy, introduced in chapter 19, of adding an ester group and disconnecting to two identical molecules, here the aldehydoester **37** still having the 1,5-diCO relationship, and we can use a malonate to make sure we get conjugate addition to acrolein.

The synthesis is straightforward except for the Krapcho method (NaCl in wet DMSO) of decarboxylating one ester in a malonate **39** without hydrolysing the other. After the condensation this was used again to give the ketone **34** and finally the reductive amination used NH$_4$OAc and NaB(CN)H$_3$ as discussed in chapter 4.

Enamines

We met enamines as specific enol equivalents in the last chapter and they are particularly good at conjugate addition. The pyrrolidine enamine from cyclohexanone **41** adds to acrylic esters **42** in conjugate fashion and the first-formed product **43** gives the enamine **44** by proton exchange.[4] Acid hydrolysis via the imine salt **45** gives the 1,5-dicarbonyl compound **46**.

The unsymmetrical diketones **47** were needed for photochemical experiments. The better 1,5-diCO disconnection **47a** is at the branchpoint. The enone **49** can be made by the Mannich reaction (chapter 20).

An enamine **52** was chosen for the synthon **48**, using morpholine **51** as the secondary amine.[5]

Among the best specific enol equivalents for Michael addition are silyl enol ethers that are rather beyond the scope of this book but are treated in detail in *Strategy and Control*. So the silyl enol ether **54** of the ester **53** adds to the enone **55** with Lewis acid catalysis to give a reasonable yield of the ketoester **56** considering that two quaternary centres are joined together.[6]

Michael Acceptors Good at Conjugate Addition

Compounds that Resist Direct Attack

Unsaturated nitro compound and nitriles do not usually suffer nucleophilic attack by enols or enolates and both are good at conjugate addition. The addition of the morpholine enamine **57** of cyclohexanone to **58** demonstrates that the nitro group is more effective than the ester at promoting conjugate addition.[7]

alkenyl nitro compounds   alkenyl nitriles

Less Electrophilic Carbonyl Compounds

We have already established that RCHO and RCOCl are poor at conjugate addition while ketones and esters are better. An extreme example is the amide **61** that does conjugate addition even with the lithium enolate **60** of cyclohexanone.[8]

Compounds Activated Towards Conjugate Additions

If the electrophilic end of the alkene is unsubstituted, it is particularly prone to conjugate additions. Examples include *exo*-methylene lactones **63**, ketones **64** and vinyl ketones **65** that are often used as the Mannich bases **66** since the free vinyl ketones tend to dimerise. So **65**; R = Me gives the dimer **67**.

| 63 | 64 | 65 | 66 | 67 |

Compounds with a Removable Activating Group in the α-Position

Electron-withdrawing groups that can be added to the α-position and removed easily after the new C–C bond is formed, promote conjugate addition. They are activating rather than protecting groups. So the CO_2R group **68** can be removed by hydrolysis and decarboxylation, sulfur-based groups **69** and **70** by reduction with Raney Ni or amalgams, $SiMe_3$ **71** by fluoride ion and Br **72** by zinc.

| CO₂R | SPh | O=SPh | SiMe₃ | Br |
| 68 | 69 | 70 | 71 | 72 |

Unsaturated Ketones

It may comfort you to know that most α,β-unsaturated ketones will do conjugate addition if the enol(ate) equivalent is carefully chosen.

The Robinson Annelation

Combining aldol and Michael reactions in one sequence is very powerful, particularly if one of the reactions is a cyclisation. The Robinson annelation[9] makes new rings in compounds like **73** that were needed to synthesise steroids. Disconnection of the enone reveals triketone **74** having 1,3- and 1,5-dicarbonyl relationships. The 1,3-disconnection would not remove any carbon atoms but the 1,5-disconnection at the branchpoint gives a symmetrical β-diketone that should be good at conjugate addition.

| 73 | 74 | 65; R = Me | 75 |

The synthesis can be done in stages under very mild conditions. The conjugate addition happens simply in water, as with **23**. Amines catalyse the cyclisation and the dehydration of **77** is catalysed by acid.[10] Triketone **74** can be made in one pot if KOH and MeOH are used with an excess of butanone **65**; R = Me. Pyrrolidine then catalyses the cyclisation and dehydration.[11]

Alternatively, Mannich salts[12] with NaOEt or Mannich bases[13] with pyridine and HCl can be used. The intermediate **77**, not usually isolated, has a *cis* ring junction, as expected from the intramolecular reaction.[14]

The new ring need not be fused to an old one and simple cyclohexenones can be made by Robinson annelation usually with the addition of a CO$_2$Et activating group. In the disconnection of cyclohexenone **78**, you could add the CO$_2$Et to form **79** before the second disconnection, as we have done, after the second disconnection or while writing out the synthesis.

Chalcones such as **80** are very easily made by an aldol reaction between acetophenone and benzaldehyde: conjugate addition of the enolate of **81** and cyclisation occur all in the same reaction.[15] The ester **82** is formed as a mixture of diastereomers in high yield: hydrolysis and decarboxylation give **78**.

Closely related to the Robinson annelation is the sequence of conjugate addition and acylation used to make 'dimedone' **83**. Either disconnection of the 1,5-dicarbonyl compound **84** is good but we prefer **84**a as the enone **85** is the aldol dimer of acetone (chapter 19) and is readily available.

The synthesis using malonate is a one-step process with NaOEt in EtOH followed by hydrolysis and decarboxylation in the usual way to give dimedone **83** in 67–85% yield.[16]

Heterocycles Made from 1,5-diCarbonyl Compounds

A family of calcium channel antagonists based on the general structure **88** is widely used to combat high blood pressure. Disconnecting the structural C–N bonds we discover a symmetrical 1,5-diketone **89** so disconnection of either appropriate bond gives the same starting materials: and enone **90** and an acetoacetate ester **91**. One of the first was nifedipine[17] **88**; R = Me, Ar = *o*-nitrophenyl.

The enone **90** is an aldol product from an aromatic aldehyde and the same acetoacetate **91**, raising the possibility that all these reactions might occur at once. We have rediscovered the Hantzsch pyridine synthesis.[18] The three components are reacted with ammonia (often NH_4OH or NH_4OAc) to give **88** in one step.[19]

References

1. J.-F. Lavalee and P. Deslongchamps, *Tetrahedron Lett.*, 1988, **29**, 6033.
2. P. D. Bartlett and G. F. Woods, *J. Am. Chem. Soc.*, 1940, **62**, 2933.
3. R. V. Stevens and A. W. M. Lee, *J. Am. Chem. Soc.*, 1979, **101**, 7032.
4. G. Stork, A. Brizzolara, H. Landesman, J. Szmuszkovicz and R. Terrell, *J. Am. Chem. Soc.*, 1963, **85**, 207.
5. J. P. Bays, M. V. Encinas, R. D. Small and J. C. Sciano, *J. Am. Chem. Soc.*, 1980, **102**, 727.
6. K. Saigo, M. Osaki and T. Mukaiyama, *Chem. Lett.*, 1976, 163.
7. J. W. Patterson and J. McMurry, *J. Chem. Soc., Chem. Commun.*, 1971, 488.
8. K. K. Mahalanabis, Z. Mahdavi-Damghani and V. Snieckus, *Tetrahedron Lett.*, 1980, **21**, 4823.
9. M. E. Jung, *Tetrahedron*, 1976, **32**, 3.
10. Z. G. Hajos and D. R. Parrish, *J. Org. Chem.*, 1974, **39**, 1612; 1615.
11. S. Ramachandran and M. S. Newman, *Org. Synth. Coll.*, 1973, **5**, 486.
12. P. Wieland and K. Miescher, *Helv. Chim. Acta*, 1950, **33**, 2215.
13. S. Swaminathan and M. S. Newman, *Tetrahedron*, 1958, **2**, 88.
14. T. A. Spencer, H. S. Neel, D. C. Ward and K. L. Williamson, *J. Org. Chem.*, 1966, **31**, 434;   K. L. Williamson, L. R. Sloan, T. Howell and T. A. Spencer, *J. Org. Chem.*, 1966, **31**, 436.
15. R. Connor and D. B. Andrews, *J. Am. Chem. Soc.*, 1934, **56**, 2713.
16. R. L. Shriner and H. R. Todd., *Org. Synth. Coll.*, 1943, **2**, 200.
17. F. Bossert, H. Meyer and E. Wehninger, *Angew. Chem. Int. Ed.*, 1981, **20**, 762.

18. U. Eisner and J. Kuthan, *Chem. Rev.*, 1972, **72**,1; D. M. Stout and A. I. Meyers, *Chem. Rev.*, 1982, **82**, 223.

19. A. Singer and S. M. McElvain, *Org. Synth. Coll.*, 1943, **2**, 214; B. Loev, M. M. Goodman, K. M. Snader, R. Tedeschi and E. Macko, *J. Med. Chem.*, 1974, **17**, 956.

22 Strategy X: Aliphatic Nitro Compounds in Synthesis

Background Needed for this Chapter Reference to Clayden, *Organic Chemistry:* Chapter 26: Alkylation of Enolates.

In chapter 21 we mentioned nitro compounds as promoters of conjugate addition: they also stabilise anions strongly but do not usually act as electrophiles so that self-condensation is not found with nitro compounds. The nitro group is more than twice as good as a carbonyl group at stabilising an 'enolate' anion. Nitromethane ($pK_a \sim 10$) **1** has a lower pK_a than malonates **4** ($pK_a \sim 13$). In fact it dissolves in aqueous NaOH as the 'enolate' anion **3** formed in a way **2** that looks like enolate anion formation.

Few aliphatic nitro compounds are wanted as target molecules in their own right but the nitro group is important in synthesis because it can be converted into two functional groups in great demand: amines **7**, by reduction, and ketones **5**, by various forms of hydrolysis.

The reduction is straightforward: the N-O bond is weak and is reduced by catalytic hydrogenation but the 'hydrolysis' needs some comments. Early and violent methods included the Nef reaction[1]—the hydrolysis of the 'enol' form **8** in strong acid, probably via the intermediate **10** with liberation of nitrous oxide N_2O.

Organic Synthesis: The Disconnection Approach. Second Edition   Stuart Warren and Paul Wyatt
© 2008 John Wiley & Sons, Ltd

Strangely enough other methods use either oxidation or reduction. The anion **8** can be oxidised at the C=N double bond by ozone[2] or $KMnO_4$ (permanganate).[3] On the other hand, the imine formed by reduction of the N–O bonds, can be hydrolysed to the ketone. It seemed that $TiCl_3$ was the solution to these problems as the McMurry reaction[4] gives excellent yields of ketones. But the recent surge in price of all Ti(III) salts has made this less attractive.

Nitro compounds can be alkylated and are good at conjugate addition (chapter 21) so the products of these reactions can be used to make aldehydes, ketones and amines. A simple synthesis of octanal[5] shows that these methods can work very well indeed. Alkylation of nitromethane with bromoheptane gives the nitro-compound **11**. Formation of the anion **12** and oxidation with $KMnO_4$ gives octanal in 89% yield. This chemistry gives us the disconnection to an alkyl halide and a carbonyl anion. The anion **12** is an 'acyl anion equivalent' and we shall need these in the next chapter.

Reduction of Nitro Compounds

The sequence of alkylation followed by reduction gives an amine and the special advantage of this strategy is that it can lead to *t*-alkyl amines. The appetite suppressant **15** can be disconnected next to the tertiary centre after the amines are changed to a nitro-compound **16**. 2-Nitropropane **18** is available.

The synthesis uses alkylation by a benzylic halide and the reduction of both nitro groups is done catalytically with Raney nickel in the same step.[6]

Other groups beside nitro can be reduced in the same step. So the diamine **19**, needed for polyamine manufacture, could come from the unsaturated nitro compound **20** that would in turn come from an 'aldol' reaction between the anion of nitromethane **1** and the aldehyde **21**. This has a 1,5-diX relationship and acrylonitrile **23** is excellent at conjugate addition (chapter 21) so we can use isobutyraldehyde **24** as a starting material.

In the synthesis we should not wish to make **21** as it would cyclise and, in any case, we'd rather reduce nitrile, nitro and alkene all in the same step by catalytic hydrogenation. The very simple method used for the conjugate addition is possible only because of the slow aldol reaction of the hindered aldehyde **24**. The 'aldol' **25**, also called a Henry reaction, needs a separate dehydration step but the three functional groups in **26** are reduced in one step in good yield.[7]

The 'nitro-aldols' can also be converted to ketones. The enantiomerically pure aldehyde **27** (a protected form of glyceraldehyde) reacts with **28** to give the 'aldol' **29** as a mixture of diastereoisomers. The protecting group 'R' is the very hindered TIPS group (i-Pr)$_3$Si. Dehydration by DCC catalysed by Cu(I) gives the nitroalkene **30** as an E/Z mixture.

Reduction of **30** with the mild reducing agent Zn/HOAc at 0 °C gives the oxime **31** that can be hydrolysed directly to the ketone **32** without isolation.[8] This ketone was used in a synthesis of compactin.[9]

Nitro-alkanes are good at conjugate additions too. In a synthesis of an immunosuppressant for organ transplants, the spirocyclic amido-ketone **33** was needed. As this is a symmetrical ketone we can use the strategy of adding an ester group and then a 1,3-diCO disconnection **34** to give symmetrical **35**. You might have disconnected the amide first but whenever you do it, you should expose an even more symmetrical compound **36**. Can we use this symmetry?

If we change the amine in **36** to a nitro group **37**, three conjugate additions of methyl acrylate to nitromethane become a possibility. Though you need quick thinking to see this, all the disconnections are ones we have seen before.

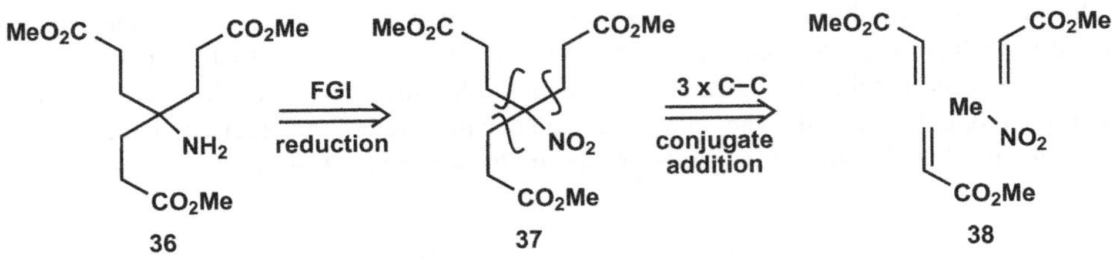

36   37   38

The synthesis is of course very short.[10] Three equivalents of methyl acrylate add to nitro-methane with catalytic (5%) DBU to give the adduct **37** and reduction leads to spontaneous cyclisation of one of the ester groups to give **35**. The rest is as planned.

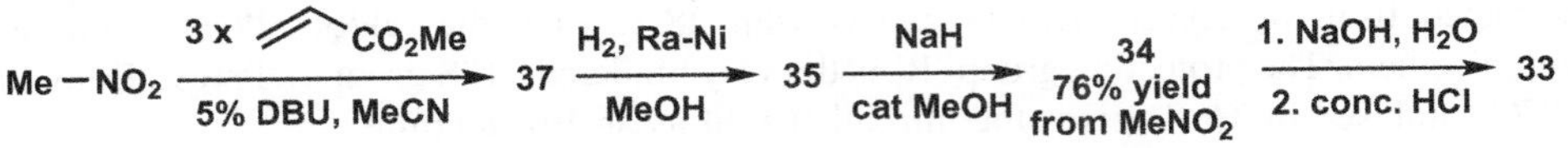

Diels-Alder Reactions

Nitroalkenes (see **30**) are easily made from nitro-alkanes and aldehydes and take part as dieno-philes in Diels-Alder reactions (chapter 17). The products can, as usual, be converted into amines or ketones. The stimulant fencamfamin **39** disconnects to the obvious Diels-Alder adduct **41** from cyclopentadiene **42** and the nitro-alkene **43**.

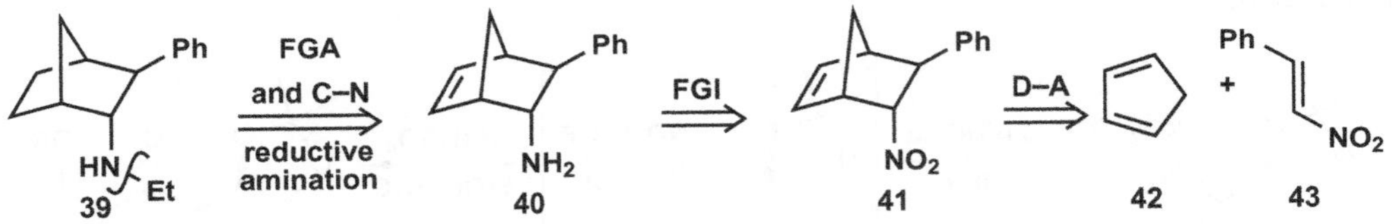

39   40   41   42   43

The synthesis starts as planned and catalytic hydrogenation reduces both the alkene and the nitro group in one step to give **44**. In the reductive amination, the imine can be formed and then hydrogenated.[11]

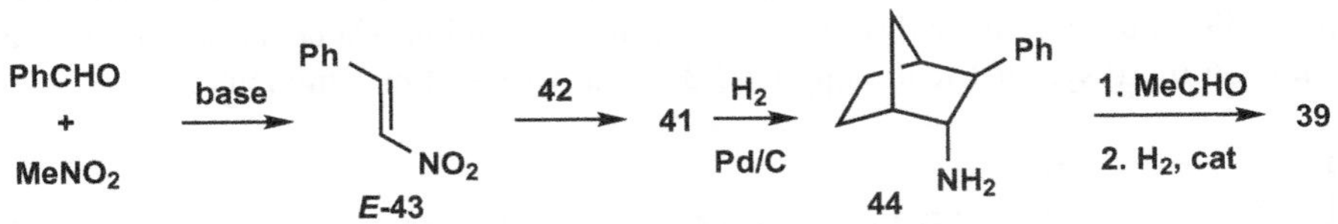

E-43   44

Now, how about making the ketone **47** by the Diels-Alder reaction? Direct disconnection (arrows on **47**) leads to a good diene **45** but an unacceptable dienophile **46**. This is a ketene and they don't do Diels-Alder reactions. You will see in chapter 33 what they can do. But if you change the ketone into a nitro group **48**, the problem disappears.

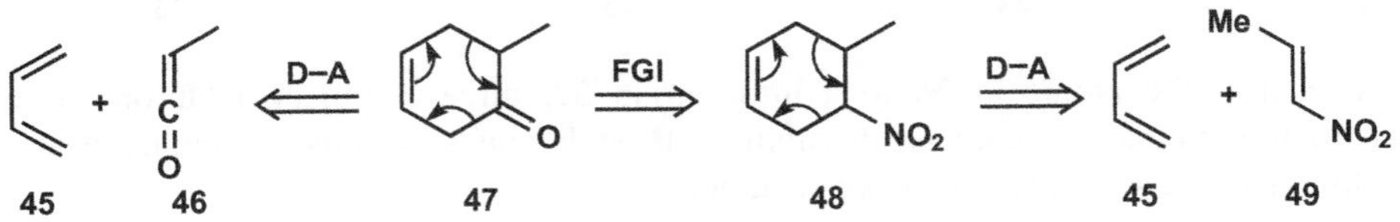

45   46   47   48   45   49

This is the work of McMurry so you can expect him to use his reagent ($TiCl_3/H_2O$) to convert the nitro group into the ketone **47**. The stereochemistry of the Diels-Alder adduct **48** is of no interest as both diastereomers give[2] **47**.

Summary of Nitro Groups in Synthesis

The nitro group is remarkably versatile and solves otherwise difficult problems. The table is meant to help you see which synthons can be represented by nitro-compounds. Note particularly that the charged synthons all have unnatural polarity and the primary enamine in the Diels-Alder entry could not be made without protection of the amine.

TABLE 22.1 Synthons represented by the nitro group

Reaction	Example	Synthon Represented	
		if reduced	if ketone made
Alkylation			
Nitro 'aldol'			
Conjugate addition			
Conjugate addition with nitro alkenes			
Diels-Alder			

References

1. W. E. Noland, *Chem. Rev.*, 1955, **55**, 137.
2. J. E. McMurry, J. Melton and H. Padgett, *J. Org. Chem.*, 1974, **39**, 259.
3. N. Kornblum, A. S. Erikson, W. J. Kelly and B. Henggeler, *J. Org. Chem.*, 1982, **47**, 4534.

4. J. E. McMurry, *Acc. Chem. Res.*, 1974, **7**, 281;   J. E. McMurry and J. Melton, *J. Org. Chem.*, 1973, **38**, 4367.

5. Vogel, page 600.

6. H. B. Hass, E. J. Berry and M. L. Bender, *J. Am. Chem. Soc.*,1949, **71**, 2290; G. B. Bachmann, H. B. Hass and G. O. Platau, *Ibid.*, 1954, **76**, 3972.

7. G. Poidevin, P. Foy and T. Rull, *Bull. Soc. Chim. Fr.*, 1979, II-196.

8. Reduction to oxime: H. H. Baer and W. Rank, *Can. J. Chem.*, 1969, **47**, 145.

9. A. K. Ghosh and H. Lei, *J. Org. Chem.*, 2002, **67**, 8783.

10. T. Kan, T. Fujimoto, S. Ieda, Y. Asoh, H. Kitaoka and T. Fukuyama, *Org. Lett.*, 2004, **6**, 2729.

11. G. I. Poos, J. Kleis, R. R. Wittekind and J. D. Rosenau, *J. Org. Chem.*, 1961, **26**, 4898;   J. Thesing, G. Seitz, R. Hotovy and S. Sommer, *Ger. Pat.*, 1,110,159, (1961); *Chem. Abstr.*, 1961, **56**, 2352h.

23 Two-Group Disconnections IV: 1,2-Difunctionalised Compounds

Background Needed for this Chapter References to Clayden, *Organic Chemistry:* Chapter 20: Electrophilic Addition to Alkenes.

In chapters 19 (1,3-diCO) and 21 (1,5-diCO) we were able to use an enol(ate) as the carbon nucleophile when we made our disconnection of a bond between the two carbonyl groups. Now we have moved to the even-numbered relationship 1,2-diCO this is not possible. In the simple cases of a 1,2-diketone **1** or an α-hydroxy-ketone **4**, there is only one C–C bond between the functionalised carbons so, while we can use an acid derivative **3** or an aldehyde **5** for one half of the molecule, we are forced to use a synthon of unnatural polarity, the acyl anion **2** for the other half. We shall start this chapter with a look at acyl anion equivalents (d^1 reagents) and progress to alternative strategies that avoid rather than solve the problem.

Acyl Anion Equivalents

The simplest reagent for an acyl anion is cyanide ion, one of the few genuine carbanions. After addition to an aldehyde, say, the resulting cyanohydrin **7** can be converted into a range of compounds **6** and **8–10**. The cyanide ion represents the synthons shown in frames next to each product.

Organic Synthesis: The Disconnection Approach. Second Edition   Stuart Warren and Paul Wyatt
© 2008 John Wiley & Sons, Ltd

Despite this versatility, cyanide adds only one carbon atom of course and we need other more general acyl anion equivalents. In chapter 16 we saw how acetylenes can give rise to ketones by hydration. A very simple example is the hydroxy-ketone **11** that could come from the acetylenic alcohol **12** by hydration and hence from acetone with the anion of acetylene acting as the acyl anion equivalent.

The sodium salt of acetylene adds to acetone and alcohol **12** can by hydrated in acid with Hg(II) catalysis.[1]

The cyclohexenone **13** was needed for a synthesis of the boll weevil hormone grandisol. Disconnection with Robinson annelation (chapter 21) in mind gives the rather unstable looking enone **15**. No doubt a Mannich method could be used (**16**; R = NR$_2$) but any leaving group X in **16** will do.

If we put X = OMe we have a skeleton that could be made by hydration of the symmetrical acetylene **17** and the ethers put in by alkylation of the diol **18**. The diol is available because it is easily made from acetylene and formaldehyde.

Alkylation with dimethylsulfate and base gave the diether and the usual hydration with Hg(II) gave the ketone[2] **16**; X = OMe.

The Robinson annelation can be carefully completed by the preparation of the enone **15** in acid and combination with the enamine[3] of *i*-PrCHO or more simply by combining **16**; X = OMe with *i*-PrCHO in base.[4]

There are many other acyl anion equivalents that are dealt with in detail in *Strategy and Control*.[5] An example of this general approach is phenaglycodol **19** used in the treatment of mild epilepsy. This 1,2-diol could be made in many ways but disconnection of two methyl groups reveals an α-hydroxy-ester **20** that could be made by addition of cyanide to the ketone **21**.

The ketone comes from a Friedel-Crafts reaction, the cyanohydrin was hydrolysed in two stages via the amide **23** and an excess of MeMgI on the ethyl ester **20**; R = Et gave the diol[6] **19**.

Other Acyl Anion Equivalents

Cyanide (one carbon) and acetylene (two carbons) are limited and other acyl anion equivalents are more versatile. Dithians are thioacetals of aldehydes that can be deprotonated between the two sulfur atoms by strong bases such as BuLi. Reaction with a second aldehyde gives **27** and hydrolysis of the thioacetal by acid, usually catalysed by Cu(II) or Hg(II), gives the α-hydroxyketone **4**. The disconnection is that shown on diagram **4** and the lithium derivative **26** acts as the acyl anion **2**. Unlike previous methods, R^1 does not have to be H or Me.

When Knight and Pattenden[7] wanted to make a group of natural products from lichens, including 'vulpinic acid' used by Eskimos to poison wolves, they needed the ketoacid **30** and could have used, say, cyanide for the synthon **29** on some acylating agent but chose instead to use a dithian as a reagent for **31** and carbon dioxide as **32**.

The dithian **33** was made from the available aldehyde, acylated with CO_2 to give **34** and the dithian hydrolysed with Cu(II) catalysis to give the α-ketoacid **30** all in excellent yield. Dithians are easy to make, stable and easy to use, but the deprotection can be tricky.

An alternative was used by Baldwin in the work that led to his famous rules for cyclisation.[8] He needed to study the cyclisation of the hydroxy-enone **36** and an obvious aldol disconnection led back to the α-hydroxy-ketone **37**. The same disconnection requires the addition of an acyl anion equivalent **39** to cyclohexanone and Baldwin chose the lithium derivative **40** of a vinyl ether.

These compounds are the opposite of the dithians: much easier to hydrolyse but more difficult to make and use. *t*-BuLi was needed for the deprotonation and the rest of the synthesis was straightforward. Note the high yield in the deprotection and that the aldol is unambiguous: only the ketone **37** can enolise and the aldehyde is more electrophilic.

Methods from Alkenes

Seeing the diol **19** you might first have though of hydroxylating an alkene and, if so, that was a good idea. Alkenes react with many electrophiles to give 1,2-difunctionalised compounds. So the 1,2-diol **43**, at a more reduced oxidation level than we have considered so far, would easily come from the alkene **44** by dihydroxylation with OsO_4. No disconnection so far, but we might first think of a Wittig reaction, and then the disconnection is across the alkene, equivalent to the bond between the OH groups in **43**. Further disconnection reveals that we would be coupling an alkyl halide **47** with an aldehyde **45**. There are of course many other ways to make alkenes (chapter 15) that would use different disconnections.

Epoxides give rise to many 1,2-difunctionalised compounds such as **48** with control over stereochemistry. Reactions of the epoxide **49** from **44** give the *anti* stereochemistry in **48** in contrast to the *syn* stereochemistry in **43**. Other compounds made from alkenes include 1,2-bromides and bromohydrins from reaction with bromine alone or bromine and water.

An example is the study by Lambert[9] of the influence of aromatic rings and neighbouring electron-withdrawing groups on S_N2 reactions. He needed the *bis*-tosylate **50**. This comes from the diol **51** and now he had a choice. He could epoxidise an *E*-alkene or dihydroxylate a *Z*-alkene. He chose the latter as *Z*-**52** could be made by a Wittig reaction.

He made both starting materials from esters of the corresponding aryl acetic acids **53** and **56** by reduction and substitution (in the case of the phosphonium salt). This gave added flexibility as either component can be used as the phosphonium salt **55** or the aldehyde **54**. He used the unusual reducing agent REDAL [$NaAlH_2(OCH_2CH_2OMe)_2$] instead of DIBAL to make the aldehyde **54**.

PhLi gave the ylid from **55** and the Wittig reaction with **54** did indeed give *Z*-**52**. These days we should probably use catalytic OsO_4 for the dihydroxylation but his mixtures [1. AgOAc, I_2, HOAc, H_2O, 2. KOH, EtOH] also gave the diol **51** and TsCl in pyridine gave the *bis* tosylate **50**. This chemistry is explained in the workbook.

α-Functionalisation of Carbonyl Compounds

We used this strategy in chapter 6 under two-group C–X disconnections where bromination of ketones was the usual functionalisation. More relevant here are conversions of carbonyl compounds into 1,2-dicarbonyl compounds by reaction with selenium dioxide SeO_2 or by nitrosation. So acetophenone **57** gives the ketoaldehyde[10] **58** with SeO_2. These 1,2-dicarbonyl compounds are unstable but the crystalline hydrate **59** is stable and **58** can be reformed on heating. Since aromatic ketones such as **57** would certainly be made by a Friedel-Crafts reaction the disconnection **58a** is not between the two carbonyl groups and offers an alternative strategy.

Nitrosation of the enol of **60** in acid solution and tautomerisation of the nitroso compound **61** gives the oxime **62**. Hydrolysis of the oxime **62** gives the diketone **63**.

Examples of α-Functionalisation of Carbonyl Compounds

Metaproterenol **64** is an adrenaline analogue used as a bronchodilator.[11] The amine might be inserted by reductive amination on the aldehyde **65** and this might be made by α-functionalisation of the available ketone **66**.

The phenols need to be protected as their methyl ethers **67** and functionalisation by SeO_2, as described earlier in this chapter, gives the keto-aldehyde **68**. To get **65** we should have to reduce the ketone in the presence of the aldehyde but the workers at Boehringer discovered a shortcut: reductive amination using hydrogenation reduced both the imine (from *i*-PrNH₂ and the aldehyde) and the ketone to give **69** and hence, by deprotection, metaproterenol **64**. Notice that the aldehyde in **68** is more electrophilic than the conjugated ketone so it forms the imine needed for reductive amination.

The triester **70** was needed to study pericyclic reactions with electron-rich (a) and electron poor (b) alkenes.[12] The α,β-unsaturated carbonyl disconnection reveals an enolisable ester **72** (X is some activating group such as CO_2R) and a very electrophilic keto-diester **71**. The synthesis of the allyl ester **72** is all right but the tricarbonyl compound **71** with *two* 1,2-diCO relationships, is a challenge.

Nitrosation of malonate **73** with N_2O_4 (and hydrolysis of the oxime) gave **71**. A Wittig reaction was chosen for the coupling which makes sure that no reaction occurs at the esters.

Strategy of Available Starting Materials

Since the 1,2-relationship is difficult to set up by making the bond between the two functionalised carbons, we may choose not to do what we did in the synthesis of **50** and **71**. We can instead buy the 1,2-relationship by using a starting material that already contains it. A simple example would be the diol **74** in which we might recognise the bones of lactic acid **75**. We therefore need to disconnect both phenyl groups, **74a** thinking of the addition of PhMgBr or PhLi to an ester of lactic acid **76** with the remaining OH group protected.

In practice, lactic acid dimerises on heating to give the double lactone **77** with both OH groups protected and treatment of this 'lactide' **77** with an excess of PhMgBr gives the diol[13] **74**.

Available compounds with a 1,2-relationship include many simple ones **78–90** whose trivial names may help you find them in suppliers' catalogues. The amino acids **83** are the constituents of proteins and are available with R = alkyl, aryl, and various functionalised groups.[14]

78; oxalic acid
79; glyoxal as aqueous solution
80; glyoxylic acid (as .H_2O)
81; glycollic acid
82; pyruvic acid
75; lactic acid
83; amino acids

84; butane dione
85; chloracetyl chloride
86; benzoin
87; benzil
88; glycol
89; ethan -olamine
90; ethylene diamine

We have also made useful starting materials in this chapter such as **11** and **50**. When the true structure of bullatenone **91** was discovered, the synthesis needed enone **92**. This is an enol ether and can be made from the aldehyde **93** and hence from **11** that we made earlier in the chapter.

91   92   enol ether   93   1,3-diCO HCO_2Et +   11

No control is needed in the first step as only the ketone **11** can enolise and the formate ester HCO_2Et is more electrophilic.[15] The product is isolated as the hemiacetal **94** that dehydrates to **91** on distillation.

11   NaH / HCO_2Et 93   94   distil   92; 50% yield

The Benzoin Condensation

If an α-hydroxyketone **95** is symmetrical the disconnection we started with **4** offers an intriguing possibility: could the acyl anion **96** be made from the aldehyde **97**? The answer is 'yes' providing R has no enolisable hydrogens, especially if it is aromatic. So, treating benzaldehyde with catalytic cyanide ion gives **98** in one pot.[16]

95   1,2-diCO   96   97   97; R = Ph   cat. NaCN / EtOH   98; 90% yield

Cyanide adds to the aldehyde **99** forming **100** which exchanges a proton to give a cyanide-stabilised anion that adds **101** to a second molecule of benzaldehyde. Exchange of a proton allows the release of the cyanide **102** so that it can be used again.

This reaction – the benzoin condensation[17] – is the nearest we have come to realising the simplest strategy of acyl anion and carbonyl electrophile in one step. One important group of reactions that make 1,2-difunctionalised compounds is the subject of the next chapter on radical reactions. A more modern version of this reaction, not needing cyanide, is described in chapter 39.

References

1. M. A. Ansell, W. J. Hickinbottom and A. A. Hyatt, *J. Chem. Soc.*, 1955, 1592.
2. G. F. Hennion and F. P. Kupiecki, *J. Org. Chem.*, 1953, **18**, 1601.
3. G. L. Lange, D. J. Wallace and S. So, *J. Org. Chem.*, 1979, **44**, 3066.
4. E. Wenkert, N. F. Golob and R. A. J. Smith, *J. Org. Chem.*, 1973, **38**, 4068; E. Wenkert, D. A. Berges and N. F. Golob, *J. Am. Chem. Soc.*, 1978, **100**, 1263.
5. *Strategy and Control*, chapter 14.
6. *Drug Synthesis* page 219–220; C. H. Boehringer Sohn, *Belg. Pat.*, 1961, 611, 502; *Chem. Abstr.*, 1962, **57**, 13678i.
7. D. W. Knight and G. Pattenden, *J. Chem. Soc., Perkin Trans. 1*, 1970, 84.
8. J. E. Baldwin, J. Cutting, W. Dupont, L. Kruse, L. Silberman and R. C. Thomas *J. Chem. Soc., Chem. Commun.*, 1976, 736; G. A. Höfle and O. W. Lever, *J. Am. Chem. Soc.*, 1974, **96**, 7126.
9. J. B. Lambert, H. W. Mark and E. S. Magyar, *J. Am. Chem. Soc.*, 1977, **99**, 3059; J. B. Lambert, H. W. Mark, A. G. Holcombe and E. S. Magyar, *Acc. Chem. Res.*, 1979, **12**, 321.
10. *Vogel*, page 627.
11. *Drug Synthesis*, pages 64–65; C. H. Boehringer Sohn, *Belg. Pat.*, 1961, 611502; *Chem. Abstr.*, 1962, **57**, 13678i.
12. B. B. Snider, D. M. Roush and T. A. Killinger, *J. Am. Chem. Soc.*, 1979, **101**, 6023.
13. M. S. Kharasch and O. Reinmuth, *Grignard Reactions of non-Metallic Substances*, Prentice-Hall, New York, 1954, page 688.
14. Clayden, *Organic Chemistry*, chapter 49.
15. P. Margaretha, *Tetrahedron Lett.*, 1971, 4891; A. B. Smith and P. J. Jerris, *Synth. Commun.*, 1978, **8**, 421; S. W. Baldwin and M. T. Crimmins, *Tetrahedron Lett.*, 1978, 4197; *J. Am. Chem. Soc.*, 1980, **102**, 1198.
16. *Vogel*, page 1044.
17. W. S. Ide and J. S. Buck, *Org. React.*, 1948, **4**, 269; A. Hassner and K. M. L. Rai, *Comp. Org. Synth.*, **1**, 542.

24 Strategy XI: Radical Reactions in Synthesis

Background Needed for this Chapter Reference to Clayden, *Organic Chemistry:* Chapter 39: Radical Reactions.

We have so far discussed only ionic and pericyclic reactions and rightly so for they are more important in synthesis than the third type: radical reactions.[1] However, some radical reactions are useful and it is appropriate to put them here as many of them lead to 1,2-difunctional compounds.

Functionalisation of Allylic and Benzylic Carbons[2]

Ionic routes to allylic **4** and benzylic **6** alcohols include reduction of the ketones **3** and **5** as these are easily made by aldol reactions and Friedel-Crafts acylation. The alcohols can be converted into electrophiles by tosylation or conversion into bromides.

Radical reactions give direct routes to allylic and benzylic halides from hydrocarbons and these reactions add functionality to previously unfunctionalised carbon atoms. The reagent is a bromine radical, that is, an atom of bromine having an unpaired electron. Bromine molecules can be decomposed by light with homolytic cleavage of the weak Br–Br bond **7** and the bromine radicals can then abstract hydrogen atoms **8** from the activated positions next to an alkene or a benzene ring so that the intermediate carbon-centred radical **9** is stabilised by conjugation. The last step is the capture of a bromine atom from a molecule of bromine and it is important that this also releases a new bromine radical to start the reaction again. This is a radical chain process.

The reaction may not even need light: just refluxing *p*-nitrotoluene **11** with bromine in petrol ether gives a moderate yield of the benzylic bromide **12**. Benzyl chloride **14** can be

---

Organic Synthesis: The Disconnection Approach. Second Edition   Stuart Warren and Paul Wyatt
© 2008 John Wiley & Sons, Ltd

made direct from toluene **13** with sulfuryl chloride activated by the radical initiator dibenzoyl peroxide.[3]

Bromine is often replaced by NBS **15** in these reactions, as in the double bromination[4] of **16**. NBS provides a low concentration of bromine and initiates the reaction by thermal cleavage of the weak N–Br bond (see workbook for details).

Allylic bromination normally uses NBS as bromine itself would add to the alkene. Thus cyclohexene gives the dibromide **18** with Br_2 but the allylic bromide with NBS. Bromine radicals abstract one of the marked H atoms from **19** and the intermediate allylic radical **21** is delocalised so we don't know which end of the allylic system[5] ends up attached to Br.

A Synthesis of Biotin

When Confalone was planning his synthesis[6] of biotin **22**, the coenzyme that carries CO_2 round the body, he observed the continuous chain of nine carbon atoms and wondered if seven of them could come from a cycloheptene. Notice that the detached two are joined to the right heteroatoms: C-8 to N and C-9 to S.

22; biotin

23; key intermediate in Confalone's synthesis of biotin

One reason for the nitro group was to use conjugate addition **23a** of the thiol **25** to nitroethylene **24**. Now it is clear that allylic functionalisation of cycloheptene **27** is the first step. In the retrosynthetic analysis we call this FGI (Functional Group Interconversion) as it cannot occur without an allylic or benzylic substituent.

NBS was used for the allylic bromination and protection was needed for the thiol nucleophile to avoid over-reaction (chapter 5). The reactive nitroalkene **24** was introduced by elimination from 2-nitroethyl acetate.

Carbon–Carbon Bond-Forming Reactions

Some radical reactions are used industrially on a large scale including radical-induced polymerisations but these are beyond the scope of this book. A few simple molecules are also made this way including the diene **29** needed for the manufacture of pyrethroid insecticides. As the molecule is symmetrical, disconnection in the middle gives two identical halves providing we make them radicals and not cations or anions. The reaction is carried out at ICI by mixing butene **31** and the allylic chloride **32** at very high temperature.[7]

The Pinacol Reaction

The same idea allows us to make symmetrical 1,2-diols **33** by the pinacol reaction. Again we avoid cations and anions by making both halves radicals. These are generated by addition of electrons from metals to aldehydes and ketones. So an electron from sodium adds to acetone to give a radical anion **35** that might dimerise to give **33**.

One popular way to perform the reaction is to use magnesium amalgam as this avoids the formation of anions: indeed the magnesium atom holds the two radicals together **36** so that the dimer

37 is formed intramolecularly. Hydrolysis gives the stable hexahydrate **38** and, if necessary, this can be dehydrated to pinacol **33** itself. Pinacol is the old name for this diol but the name is now mainly used for the reaction.

Other electron-donating systems include zinc dust and Me_3SiCl, that gives the silyl ethers[8] such as **40**, or samarium iodide that gives similar dimers of aromatic aldehydes in excellent yield.[9] These reactions usually favour the *anti*-isomers **40** and **42**.

Example: Dienoestrol

The synthetic oestrogen dienoestrol **43** might be made[10] by dehydration of the symmetrical diol **44** and hence by pinacol dimerisation of the ketone **45**. Successful pinacol with magnesium metal gave **44** that could be dehydrated with AcCl in Ac_2O.

The Acyloin Reaction

The acyloin reaction is a similar radical dimerisation but at the ester oxidation level.[11] At first it looks just like a pinacol: electrons are added to the carbonyl groups to give **47**, the radicals combine to give a new C–C bond and the ethoxide groups are lost **48** to give the 1,2-diketone **49**.

That is just the start. If the diester **46** can accept electrons, the α-diketone **49** will do so more avidly giving a new diradical **50** that forms a C–C π-bond **51** and, on work-up, the ene-diol **52** that is the less stable tautomer of the α-hydroxyketone **53**, also called an 'acyloin' and hence the name of the reaction.

However, even this is not the end as, if the reaction is done in this way, those two molecules of ethoxide released from **48** catalyse an intramolecular Claisen ester condensation **54** and the main product is the ketoester **55**.

The solution[12] is to carry out the reaction in the presence of Me$_3$SiCl. This does two things. The more obvious is that the enediol dianion **51** is trapped as the silyl enol ether **56**, a useful intermediate, but the more important thing is the removal of the basic ethoxide ions as the neutral silyl ether EtOSiMe$_3$.

So, if the ester is enolisable, use the Me$_3$SiCl method. If it isn't enolisable you don't need to. The α-diketones **57** were needed for the synthesis of tetronic acids. Changing the oxidation level of one ketone reveals a symmetrical acyloin **58** derived from the esters **59**.

These esters **59** are very enolisable so the silicon method must be used.[13]

By contrast, the α-hydroxyketone **61** is an acyloin derived from a diester **62** that has no α-Hs, cannot enolise, and will not need the silicon treatment. Simple C–S disconnection reveals a chloroester **63**.

You may be surprised to know that the chloroacid **65** can be made from available pivalic acid **63** by photochemical chlorination. This is again a radical reaction, chlorine radicals abstracting one of the *nine* hydrogens from the *t*-butyl group as there are no others. Though the chloride in **65** is rather unreactive, it combines well with sulfide anions and the acyloin goes in good yield without any silicon.[14]

Making 1,2-Difunctionalised Compounds

We end this chapter with a review of ways one might approach making diaryldiketones such as **16**. It helps that α-hydroxyketone **67** or diol **68** can be oxidised to the diketone[15] **16**.

To put you out of your misery, the original authors[4] made **67** by a benzoin condensation[16] and oxidised it to the dione[17] **16**.

But how else might they have done it? Perhaps equally obvious is to use the pinacol reaction to make the diol **68** and oxidise both alcohols. We shall now broaden the discussion by writing Ar for the substituents as many of the reactions have not in fact been done with *o*-tolyl substituents. Two of the best yielding methods are Mg in the presence[18] of Me₃SiCl and samarium iodide where the colour change from blue Sm(III) to yellow Sm(II) is characteristic.[19]

The diol **70** could also be made by dihydroxylation of the alkene **73** that might be made by a Wittig reaction from the phosphonium salt **72**. As the stereochemistry of neither the alkene **73** nor the diol **70** is relevant to the synthesis of **71**, no control is discussed.

You might also have considered the addition of an acyl anion equivalent, such as the lithium derivative of the dithian **74** to ArCHO (chapter 23). There are obviously many other methods but these are the most likely.

References

1. Clayden, *Organic Chemistry*, chapter 39.
2. House, pages 478–491.
3. *Vogel*, page 864.
4. M. Verhage, D. A. Hoogwater, J. Reedijk and H. van Bekkum, *Tetrahedron Lett.*, 1979, 1267.
5. *Vogel*, page 578.
6. P. N. Confalone, E. D. Lollar, G. Pizzolato and M. R. Uskokovic, *J. Am. Chem. Soc.*, 1978, **100**, 6291; P. N. Confalone, G. Pizzolato, D. L. Confalone and M. R. Uskokovic, *Ibid.*, 1980, **102**, 1954.
7. D. Holland and D. J. Milner, *Chem. Ind. (London)*, 1979, 707.
8. J.-H. So, M.-K. Park and P. Boudjouk, *J. Org. Chem.*, 1988, **53**, 5871.
9. J. L. Namy, J. Souppe and H. B. Kagan, *Tetrahedron Lett.*, 1983, **24**, 765.
10. E. C. Dodds and R. Robinson, *Proc. Roy. Soc. Ser. B*, 1939, **127**, 148.
11. J. J. Bloomfield, D. C. Owsley and J. M. Nelke, *Org. React.*, 1976, **23**, 259.
12. K. Rühlmann, *Synthesis*, 1971, 236; J. J. Bloomfield, D. C. Owsley, C. Ainsworth and R. E. Robertson, *J. Org. Chem.*, 1975, **40**, 393.
13. P. J. Jerris, P. M. Wovkulich and A. B. Smith, *Tetrahedron Lett.*, 1979, 4517; P. Ruggli and P. Zeller, *Helv. Chim. Acta*, 1045, **28**, 741; I. Hagedorn, U. Eholzer and A. Lüttringhaus, *Chem. Ber.*, 1960, **93**, 1584.
14. N. Feeder, M. J. Ginnelly, R. V. H. Jones, S. O'Sullivan and S. Warren, *Tetrahedron Lett.*, 1994, **35**, 9095.
15. *Vogel*, page 1045.
16. W. S. Ide and J. S. Buck, *Organic React.*, 1948, **4**, 269; see tables.
17. H. Moureu, P. Chovin and R. Sabourin, *Bull. Soc. Chim. Fr.*, 1955, **22**, 1155.
18. T.-H. Chan and E. Vinokur, *Tetrahedron Lett.*, 1972, 75.
19. J. L. Namy, J. Souppe and H. B. Kagan, *Tetrahedron Lett.*, 1983, **24**, 765.

25 Two-Group Disconnections V: 1,4-Difunctionalised Compounds

Background Needed for this Chapter Reference to Clayden, *Organic Chemistry:* Chapter 26: Alkylation of Enolates.

The problem of unnatural polarity also arises in making C–C disconnections for the synthesis of 1,4-difunctionalised compounds. If we start with 1,4-diketones **1**, disconnection in the middle of the molecule gives a synthon with natural polarity **2**, represented in real life by an enolate **4**, and one of unnatural polarity, the a^2 synthon **3** represented by some reagent of the kind we met in chapter 6 such as an α-haloketone **5**.

You might think you could escape this problem by choosing the alternative disconnection **8**, but this is not so. We have more choice here: we can use the a^3 synthon **7** with natural polarity, in real life an enone, but then we shall have to use the acyl anion equivalents **6** that we met in chapter 23. Reversing the polarity gives us the naturally polarised electrophile, an a^1 synthon **9** represented by an acylating agent and the homoenolate, or d^3 synthon, **10** with unnatural polarity.

Reactions of Enol(ate)s with Reagents for a^2 Synthons

A simple example would be the keto-ester **11**. We should prefer to disconnect the bond at the branchpoint and that suggests the synthons **12** and **13**. The reagent for **13** can be the bromoester **15** but we shall need to choose our enolate equivalent carefully. It should not be too basic as the marked protons in **15** between Br and CO_2Et are rather acidic.

Organic Synthesis: The Disconnection Approach. Second Edition   Stuart Warren and Paul Wyatt
© 2008 John Wiley & Sons, Ltd

Lithium enolates and the like are not good choices but enamines are excellent. The morpholine enamine **18** is cleanly alkylated by the bromoester **15** and hydrolysis gives[1] the keto-ester **11**.

Another good choice is the easily made β-ketoester[2] **19** (compound **41** in chapter 19) as such stabilised enolate anions are not very basic.

A Synthesis of Methylenomycin

The antibiotic methylenomycin **21** was synthesised via the key intermediate **22**. Aldol disconnection reveals a diketo-ester with a 1,4-relationship and an obvious disconnection next to the branchpoint **23**. The starting materials **24** and **25** are available.

There might be some doubt about the cyclisation of **23** but the more highly substituted alkene is preferred in cyclisations under thermodynamic control. Only **22** is formed: none of the alternative **26** can be found.[3]

4-Hydroxyketones

If we do the same disconnection at the alcohol oxidation level **27** the reagent for the a^2 synthon might be an epoxide **29**. More reactive enolates such as lithium enolates are now all right.

The useful synthetic intermediate **30**, from which methyl cyclopropyl ketone can be made, comes from the 4-hydroxyketone **31**, disconnected to the enolate of acetone **32** and ethylene oxide.

Though we could use the lithium enolate of acetone, there is an advantage in using ethyl acetoacetate **24**. The intermediate **34** cyclises under the reaction conditions and the stable lactone **35** is isolated. Treatment with HBr opens the lactone and decarboxylation gives[4] **30**. This two-step sequence requires only mild conditions.

Conjugate Addition of Acyl Anion Equivalents

We have met the acyl anion or d^1 synthon in chapter 23 but for the disconnection **8** on 1,4-diketones we need a d^1 reagent that will do conjugate additions on enones such as **36**. Sadly that eliminates dithians from consideration as they are too basic (hard) and tend to add direct to carbonyl groups.

However, the simplest one-carbon d^1 reagent, cyanide ion, does do conjugate addition well so the disconnection is successful if R^1 in **1** is OH or OR.

The anti-convulsant phensuximide **37**, being an imide, comes from a dicarboxylic acid **38** with a 1,4-relationship between the two carbonyl groups. Changing one to cyanide we get back to cinnamic acid **40** as the available starting material.

In practice cyanide added only slowly to cinnamic acid so a second electron-withdrawing group (another cyanide) was needed and the cyanide **42** was used successfully.[5]

Nitroalkanes as d^1 Reagents

Nitroalkane anions are very stable and hence excellent at conjugate addition (chapter 22). Quite weak bases such as amines are enough to give the anion of **44** and hence the nitro-ketones **45**. Reduction gives the amino-ketones **46** that cyclise to give imines, reduced under the reaction conditions to pyrrolidines **47**.

A dramatic example is the synthesis of the ant pheromone monomorine **48** already discussed in chapter 8. A bold double C–N disconnection gives the amino-diketone **49** and hence, after FGI of NH_2 to NO_2, the 1,4-disconnection we have just described.

1-Nitropropane adds to the protected enone **52** with catalysis by the base tetramethylguanidine and catalytic hydrogenation completes the first reductive amination **54**. Hydrogen adds to the imine on the same side as the H atom already there.[6]

Now the acetal can be hydrolysed to give the free amino-ketone and hence the enamine **55** that is reduced by the alternative reagent (chapter 4) sodium cyanoborohydride again stereoselectively so that all three H atoms in **48** are on the same face of the bicyclic structure.

The alternative transformation of a nitroalkane into a ketone (chapter 22) is well illustrated by a one-pot process where alumina is used to catalyse the conjugate addition and an oxidative process with H_2O_2 is used to form the ketone.[7] Overall yields are good, e.g. $R^1 = Bu$, $R^2 = Et$, 90% yield.

Direct Addition of Homoenolates (d³ Reagents)

The same disconnection but of the opposite polarity requires some acylating agent for synthon **9**: this is no problem as we have various acid derivatives at our disposal. But the nucleophilic synthon **10**, a d³ synthon or homoenolate, is another matter. There is no stabilisation of the anion as drawn but if it were to cyclise to the oxyanion **56**, it would be rather more stable and there is evidence–trapping with silicon to give **57** for example–that this can occur.

The simplest way to make such a derivative is to treat a β-halocarbonyl compound **59**, easily made from the enone **58** by conjugate addition, with zinc metal to give a species that we might write as **60** but perhaps should be **61** or even **62**. Whichever is correct, it is nucleophilic at the β-carbon atom and the polarity of the enone **58** has been reversed.

One well defined compound in this series[8] is the cyclopropane **64** made[9] from chloro-propionate esters **63** and sodium in the presence of Me₃SiCl. On treatment with $ZnCl_2$, the ring is opened and a zinc homoenolate **65** with internal coordination is formed. This reacts with acylating agents to give 1,4-dicarbonyl compounds[10] such as **66**. These reactions are catalysed by Pd(0).

If aldehydes are used as the electrophiles, a kind of homoaldol reaction occurs in the presence of Me₃SiCl to give the protected γ-hydroxyesters **67** in good yield. The zinc salt (ZnCl₂ or ZnI₂), used to make **59** also acts as a Lewis acid for the reaction.

An improved version uses (*i*-PrO)₃TiCl as a catalyst: under these conditions the substituted homoenolate **69** reacts with benzaldehyde to give the lactone **70** with good stereoselectivity in favour of *syn*-**70**. The regioselectivity suggests that a cyclopropane is not involved.[11] A more extensive treatment of homoenolates is given in *Strategy and Control*.[12]

Strategy of Available Starting Materials with a 1,4-diCO Relationship

Just as with 1,2-diCO compounds, we can buy the 1,4-relationship rather than make it. Any supplier's catalogue will reveal a wide variety of such compounds: we shall mention only a few. There are simple disubstituted butanes such as the diol **71**, the diamine **72**, the dihalides **73** and succinic acid **74**.

There are some important cyclic compounds such as the lactone **75**, succinic anhydride **76**, furan **77**, and many substituted furans, particularly furfuraldehyde **78**, a by-product of breakfast cereal manufacture, and its reduction product **79** and maleic anhydride **80**.

And finally some unsaturated compounds, *cis*-butenediol **81**, butynediol **82**, fumaric acid **83** and levulininc acid **84**. And many more.

A simple but surprising example of using an available starting material is the synthesis of 1,7-difunctionalised heptan-4-ones **86** and **88**, each having *two* 1,4-diX relationships.[13] The dihaloketone **86** was made from the aldol dimer **85** of butyrolactone **75**, as described in chapter 19, where compound **85** was compound **26**. The two 1,4-diX relationships in **86** can be seen in the intermediate **85** which has lost CO_2 on its way to **86**.

The ketodiester **88** was made from furfural **78** by a Wittig reaction and treatment of the product **87** with acidic methanol.[14] The central carbon atom in **88** can be seen as an enol ether in **87**. You might like to try to draw a mechanism for this extraordinary reaction which is described in detail in the workbook.

The FGA Strategy

Butyne diol **91** is just one example of a compound that can be made by reactions of acetylene with carbonyl electrophiles (chapter 16). Reaction in turn with two different aldehydes and reduction of the triple bond leads to unsymmetrical 1,4-diols **92**.

Reaction with an aldehyde at one end and CO_2 at the other followed by reduction gives lactones. The cyclisation to the five-membered ring **95** occurs spontaneously on hydrogenation.[15]

Analysis of the final products of these two reactions suggests rather unhelpfully that acetylene is acting as the 1,2-dianion of ethane **96**. It is probably better to think of them as examples of the FGA strategy

References

1. H. Fritz and E. Stock, *Tetrahedron*, 1970, **26**, 5821.

2. R. P. Linstead and E. M. Meade, *J. Chem. Soc.*, 1934, 935.

3. J. Jernow, W. Tautz, P. Rosen and J. F. Blount, *J. Org. Chem.*, 1979, **44**, 4210.

4. T. E. Bellas, R. G. Brownlea and R. M. Silverstein, *Tetrahedron*, 1969, **25**, 5149; G. W. Cannon, R. C. Ellis and J. R. Leal, *Org. Synth. Coll.*, 1963, **4**, 597.

5. C. A. Miller and L. M. Long, *J. Am. Chem. Soc.*, 1951, **73**, 4895.

6. R. V. Stevens and A. W. M. Lee, *J. Chem. Soc., Chem. Commun.*, 1982, 102.

7. R. Ballini, M. Petrini, E. Marcantoni and G. Rosini, *Synthesis*, 1988, 231.

8. E. Nakamura and I. Shimida, *J. Am. Chem. Soc.*, 1977, **99**, 7360.

9. K. Rühlmann, *Synthesis*, 1971, 236.

10. E. Nakamura, S. Aoki, K. Sekiya, H. Oshino and I. Kuwajima, *J. Am. Chem. Soc.*, 1987, **109**, 8056.

11. H. Ochiai, T. Nishihara, Y. Tamaru and Z. Yoshida, *J. Org. Chem.*, 1988, **53**, 1343.

12. *Strategy and Control*, chapter 13.

13. O. E. Curtis, J. M. Sandri, R. E. Crocker and H. Hart, *Org. Synth. Coll.*, 1963, **4**, 278.

14. R. M. Lukes, G. I. Poos and L. H. Sarett, *J. Am. Chem. Soc.*, 1952, **74**, 1401.

15. J. P. Vigneron and V. Bloy, *Tetrahedron Lett.*, 1980, **21**, 1735; J. P. Vigneron and J. M. Blanchard, *Ibid.*, 1739; J. P. Vigneron, R. Méric and M. Dhaenens, *Ibid.*, 2057.

26 Strategy XII: Reconnection

Synthesis of 1,2- and 1,4-diCO Compounds by Oxidative C=C Cleavage

We had to be careful in chapter 25 when we wanted to add bromoketones **4** to enolates **3** to make the 1,4-dicarbonyl compound **5**. We could not use a lithium enolate because it would be too basic. No such difficulties exist in the reaction of enolates with allylic halides such as **2**. Any enol(ate) equivalent will do as there are no acidic hydrogens and allylic halides are good electrophiles for the S_N2 reaction.

But this does not make **5**: instead it makes **1** with an extra carbon atom. So what we need are oxidative methods to convert **1** into **5** by cleavage of the alkene. There are many ways to do this. The most obvious is ozone[1] that gives aldehydes from a disubstituted alkene **6** with reductive work-up or acids with oxidative work-up.

Dihydroxylation with catalytic osmium tetroxide and stoichiometric oxidant such as NMO (*N*-methylmorpholine-*N*-oxide) gives diols that can be cleaved to the same aldehydes with sodium periodate or lead tetra-acetate. It is also possible to combine either $KMnO_4$ or catalytic OsO_4 with an excess of $NaIO_4$ and complete the operation in one pot.

So, disconnection of **11** by the methods of chapter 25 gives an enolate from the ester **8** but also a bromoaldehyde **12**, a type of compound best avoided. But if we replace O by C **10** we can disconnect to allyl bromide **9**—a satisfactory electrophile in every way. You might like to call this an FGA, and we shan't quarrel with that, but read on . . .

Malonate was used in the synthesis,[2] and it makes sense to add the more reactive alkyl group last: in this case the allyl group. The oxidative cleavage was done with ozone.

These methods are compatible with a range of functional groups in spite of the vigorous nature of the reagents. In his synthesis of brevianamide B, Williams[3] ozonised the allyl group in the heterocycle **15** to give the aldehyde **16** in amazingly good yield. The allyl group had been put in as an electrophile added to an enolate.

In his synthesis of the polyether antibiotic X-506, Evans[4] needed the triol **20** with two of the OH groups protected. He added an allyl group as a nucleophile to open the epoxide **17** at the less hindered end and then protected the two OH groups as an acetal **19** before the third was introduced by ozonolysis and reduction. These two examples emphasise the versatility of the allyl group.

But what are we to call the retrosynthetic transformation of **11** into **10**? It isn't a disconnection: rather an *extra* carbon atom has been added. So we call this operation a *reconnection*: joining the target molecule back up to something to reveal the precursor. So, consider the synthesis of the *cis*-enone **21**, a structure found in insect pheromones, perfumes and flavourings. A Wittig reaction would make the *cis*-alkene from the phosphonium salt **22** but the ketoaldehyde **23** would need protection, perhaps as the acetal **24**.

The problem is how to protect the ketone rather than the aldehyde and the answer is like that for **20**: protect it when the aldehyde isn't there. Reconnection to the alkene **25** achieves this and the ketone **26** can be made by reaction of some enolate **27** with allyl bromide.

The synthesis[5] follows this pattern with ozone and reductive work-up with dimethyl sulfide ensuring that the aldehyde is not further oxidised and deprotection of **30** taking place after the Wittig reaction.

We have used a simple allyl group so far but, as the other half of the alkene is lost in the cleavage reaction, it doesn't matter much what it is. So, with α,β-unsaturated carbonyl compounds **31** and **34**, which give rise to 1,2-diCO compounds **32, 33** or **35**, it is convenient to use aldol reactions to make the alkene, and, say benzaldehyde is easier to use than formaldehyde.

The extraordinary polycyclic tetraketone staurone **36** was made[6] from **37**. The 1,2-diCO relationship in **37** is an ideal candidate for reconnection in this style.

Aldol disconnection **38a** reveals a methyl ketone with two 1,4-diCO relationships that could be made by double alkylation of some enolate **27** of acetone with ethyl bromoacetate **40**.

The synthesis used benzyl acetoacetate **41** for the double alkylation so that the benzyl ester **42** could be specifically cleaved by hydrogenation to give **39**. Condensation with unenolisable benzaldehyde is unambiguous (chapter 20) and ozone does the rest.

A Dramatic Example of FGA

The saturated hydrocarbon **43** is a pheromone from an ant. We have an obvious problem in designing a synthesis for this compound—there are no functional groups of any kind. A less obvious problem is how to relate the two stereochemical centres when they are 1,7-related.

43; (11R,17S)-11.17-dimethylhentriacontane

The solution to the second problem is to start with two enantiomerically pure compounds derived from nature and that suggests adding an alkene between the two chiral centres **44** as that could be made by a Wittig reaction from, say, **45** and **46**.

The choice of which way round to do the Wittig may appear arbitrary but it isn't. Pempo and his group[7] chose citronellal **47** and citronellol **48**, two related natural terpenes from citronella oil, as starting materials with the right stereochemistry at the one chiral centre. If you imagine a Wittig reaction between the phosphonium salt **49** and some suitable aldehyde, you will see that the central part of the molecule would be right.

47; (R)-citronellal   **48; (R)-citronellol**   **49**

However, the aldehyde cannot be citronellal as the stereochemistry would be wrong. In addition, both terminal alkenes must be oxidised away so that the rest of the molecule may be attached. So this is what they did: the left-hand half of the molecule was assembled from citronellol **48** by oxidation to the aldehyde **50** and the remaining seven carbon atoms added by a Wittig reaction. The product is mainly *Z*-**51** but this is irrelevant as the alkene will disappear. The phosphonium salt is ready for coupling to the right-hand half.

Citronellal is combined with the straight chain Grignard. This gives a 1:1 mixture of diastereoisomers—not separated—which are blocked as tosylates before the alkene is cleaved with ozone. Wittig reaction of the resulting aldehyde **56** with the phosphonium salt **53** from the other half of the molecule gives the *Z*-alkene **57**. The new alkene is also *cis* (it was *trans* in **44**) but this doesn't matter. The tosylates are reduced off with LiAlH$_4$/NaH and both alkenes hydrogenated to give our pheromone **43** in 78% yield. The only poor step is the second Wittig reaction but this accomplishes so much by joining the two halves of the molecule together so we can live with that.

The point of this synthesis is that the workers recognised that oxidative cleavage of citronellal and citronellol would give two-ended fragments that could be used to make the core of the pheromone with the right stereochemistry. We shall see in the next chapter that this reconnection strategy is vital for the synthesis of one important group of compounds: 1,6-diCOs.

References

1. D. G. Lee and T. Chen, *Comp. Org. Synth.*, 1991, **8**, 541.
2. M. T. Edgar, G. R. Pettit and T. H. Smith, *J. Org. Chem.*, 1978, **43**, 4115.
3. R. M. Williams, T. Glibka and E. Kwast, *J. Am. Chem. Soc.*, 1988, **110**, 5927.
4. D. A. Evans, S. L. Bender and J. Morris, *J. Am. Chem. Soc.*, 1988, **110**, 2506.
5. W. G. Taylor, *J. Org. Chem.*, 1979, **44**, 1020.
6. R. Mitschka and J. M. Cook, *J. Am. Chem. Soc.*, 1978, **100**, 3973.
7. D. Pempo, J. Viala, J.-L. Parrain and M. Santelli, *Tetrahedron: Asymmetry*, 1996, **7**, 1951.

27 Two-Group C–C Disconnections VI: 1,6-diCarbonyl Compounds

Background Needed for this Chapter Reference to Clayden, *Organic Chemistry:*
Chapter 37: Rearrangements (The Baeyer–Villiger Rearrangement).

We come to the last of our chapters on two-group C–C disconnections and it has been left to the
last for a good reason. If we try to start in the same way as we have with the other chapters of
this kind, with a generalised 1,6-dicarbonyl compound **1** and disconnect in the middle we might
be relieved to see an a^3 synthon **2** easily recognised as an enone in real life, but the d^3 synthon **3**,
with unnatural polarity, caused us problems in chapter 25 and now we should need a reagent for
3 that does conjugate addition. Though there are a few ways to do this, it has not been a popular
strategy. Disconnecting elsewhere is no help as the true difficulty is that the two carbonyl groups
are too far apart for this approach.

Chapter 26 introduced the strategy of reconnection and that is indeed the main strategy for
the synthesis of 1,6-diCO compounds. But there is a big difference from what we have seen
before. We no longer trim off some atoms at the far end of the alkene to be cleaved. Instead we
reconnect intramolecularly so that the marked atoms C-1 and C-6 form a ring **4** and the bond
between these atoms must be made weaker than any other bond in the molecule. Ironically we
can do this by making it a *double* bond **5**.

No atoms are lost in the cleavage reaction so that cheap cyclohexene **6** is used to make adipic
acid **7** for nylon manufacture. Any of the oxidative cleavage methods from the last chapter could
be used: Vogel[1] has a recipe using concentrated nitric acid on cyclohexanol **8** that presumably
goes by dehydration to the alkene **6** followed by oxidation, and other methods are probably used
industrially.

Organic Synthesis: The Disconnection Approach. Second Edition   Stuart Warren and Paul Wyatt
© 2008 John Wiley & Sons, Ltd

Ketoacids with a 1,6-relationship are easily made from cyclohexanone **9** by addition of an organo-lithium or Grignard reagent, dehydration of the tertiary alcohol **10** and oxidation of the cyclohexene **11** to form **12**.

The bicyclic ketone **13** was made from the simpler enone **14** that also had to be made. Aldol (α,β-unsaturated carbonyl compound—still our first choice) disconnection reveals the keto-aldehyde **15**.

This is a 1,6-dicarbonyl compound so reconnection to the cyclohexene **16** is needed. *Top tip:* write the numbers 1–6 *both* on the target molecule **15** *and* on the starting material **16** to make sure you put the substituents on the right atoms. Now FGI and removal of the methyl group reveals a simple cyclohexanone **18**.

The synthesis of **18** is discussed in the next chapter. The synthesis of **14** is a classic of its kind: the alcohol **17** is not isolated but dehydrated directly to the cyclohexene **16** and the oxidative cleavage is done by ozone. The intramolecular aldol is unambiguous as the alternative is a seven-membered ring.[2]

The Diels-Alder Route to 1,6-diCarbonyl Compounds

This last example makes it clear that we shall normally have to make the cyclohexenes we need for oxidative cleavage and one of the best routes to such compounds is the Diels-Alder reaction (Chapter 17). A generalised example would be ozonolysis of the alkene **21**, the adduct of butadiene and the enone **20**. The product **22** has a 1,6-relationship between the two carboxylic acids. Since Diels-Alder adducts have a carbonyl group outside the ring (the ketone in **21**) the cleavage products **22** also have 1,5- and 1-4-diCO relationships and would be a matter for personal judgement which of these should be disconnected instead if you choose that alternative strategy.

Heathcock required diester **23** for his synthesis of the antibiotic pentalenolactone.[3] Reconnecting the esters gives the cyclohexene **24**. We must change the two ether groups into carbonyl groups and one obvious starting material is **25**, the Diels-Alder adduct of butadiene and maleic anhydride **26**.

The synthesis followed this pattern with the ethers **24** being made immediately after the reduction of **25** and the esters made with diazomethane CH_2N_2 after oxidative cleavage.

The bicyclic double lactone **27** was used by Eschenmoser as a precursor for all four heterocyclic rings in his synthesis[4] of vitamin B_{12}. Disconnection of both lactones reveals a ketone **29**.

The ketone **29** in fact has 1,4-, 1,5- and 1,6-relationships and if we redraw it **29a** to see the 1,6-relationship clearly, being careful to get the stereochemistry right, we can reconnect to the cyclohexene **30** and hence, by Diels-Alder disconnection, find the reactive dienophile **31**. The methyl and CO$_2$H groups are *cis* in **30** and so must be *cis* in **31**.

The dienophile **31** can be disconnected in two ways as there is a carbonyl group at each end of the alkene. Controlling disconnection **31a** might be difficult as both components can enolise, but **31** is unambiguous except for the regiochemistry of enolisation of the ketone.

An aldol reaction in acid solution ensures that the more substituted enol is formed and the aldehyde is by far the most electrophilic of all the carbonyl groups. The Diels-Alder reaction gives the free acid **30** which was resolved with a chiral amine and each enantiomer used for a different part of the B$_{12}$ molecule. The slightly unusual reagent Cr(VI) was used for the alkene cleavage and acetal formation occurred spontaneously under the acidic conditions.

Cyclohexenes from Other Sources

One advantage of making 1,6-dicarbonyl compounds by cleavage of cyclic alkenes is that stereo-chemistry may be fixed in a ring that wouldn't be fixed in an open chain compound. The natural terpene (+)-2-carene **35** has a *cis* fused three-membered ring – no other arrangement is possible. Cleavage of the alkene with ozone gave the keto-aldehyde **36** with unchanged stereochemistry. This was just what McMurry needed[5] to make a series of compounds such as **37** with the same stereochemistry.

35; (+)-2-carene   **36;** 99% yield   **37**

The sequence of cyclohexene cleavage and aldol reaction on the dicarbonyl product gives ring-contracted cyclopentenes. This proved particularly valuable when Iwata[6] wanted to make subergorgic acid **41** that has three five-membered rings awkwardly joined around a quaternary carbon atom. So crowded are these compounds that they are difficult to draw clearly. Ozonolysis of the synthetic cyclohexene **38** gave the unstable dialdehyde **39** that cyclised by an aldol condensation to **40** and hence could be oxidised to **41**.

Oxidative Cleavage by the Baeyer–Villiger Reaction

Cyclohexanones may be cleaved oxidatively by peroxyacids to give seven-membered ring lactones. This is the Baeyer–Villiger rearrangement: a migration from carbon to oxygen that has the effect of inserting an oxygen atom into the ring.[7] The lactone already has a 1,6-diO relationship but this is more obvious in the hydroxy-acid **44**. This reaction is used widely in industry.[8]

The disconnection is to extrude the oxygen atom. There is a danger here. Such disconnection of both lactones **45** and **47** leads to the same cyclohexanone **46** and both cannot be right. The one that works is **45** because the migration step prefers the more highly substituted migrating group **48** as the developing positive charge on the left-hand part of the molecule can be shared more.

So when the hydroxyketone **49** was needed for a pheromone synthesis, it was made by nucleophilic displacement on the lactone **50** by an organo-lithium compound. This lactone is of the right kind (cf. **45**) to be made by a Baeyer–Villiger rearrangement from the cyclohexanone **51** and this can be made by total reduction of the phenol **52** (chapter 36).

Catalytic reduction of the phenol gave a mixture of diastereoisomers of **53** and pure **51** could be separated from the reaction mixture after oxidation. The Baeyer–Villiger and the opening of the ring with n-octyl-Li worked well.[9]

Other Approaches

There is of course no need to use reconnection if you prefer another strategy but you are advised to try disconnection first. Disconnection of the 1,3-diCO relationship in the spiro-diketone **54** reveals a 1,6-diCO compound that could no doubt be made by oxidative cleavage of **58**. But various authors[10] preferred to ignore the 1,6-diCO relationship and simply disconnect to the enolate of cyclopentanone **56** and a bromoester **57**.

We discussed making **57** from the lactone **59** in chapter 25 and we have used the ketoester **60** in chapters 19 and 25. Alkylation of **60** gives **61**, reaction with concentrated HCl gives the acid corresponding to **55** and polyphosphoric acid (PPA) catalyses the cyclisation to **54**.

References

1. *Vogel*, page 668.
2. H. O. House, C.-C. Yao and D. Vanderveer, *J. Org. Chem.*, 1979, **44**, 3031; H. O. House and M. J. Umen, *J. Org. Chem.*, 1973, **38**, 1000.
3. F. Plavac and C. H. Heathcock, *Tetrahedron Lett.*, 1979, 2115.

4. A. Eschenmoser and C. E. Winter, *Science*, 1977, **196**, 1410.

5. J. E. McMurry, *J. Org. Chem.*, 1987, **52**, 4885.

6. C. Iwata, Y. Takemoto, M. Doi and T. Imanishi, *J. Org. Chem.*, 1988, **53**, 1623.

7. C. H. Hassall, *Org. React.*, 1953, **9**, 73;   G. R. Crow, *Org. React.*, 1993, **43**, 251.

8. G.-J. ten Brink, I. W. C. E. Arends and R. A. Sheldon, *Chem. Rev.*, 2004, **104**, 4105.

9. G. Magnusson, *Tetrahedron*, 1978, **34**, 1385.

10. W. E. Bachmann and W. S. Struve, *J. Am. Chem. Soc.*, 1941, **63**, 2590;   W. Carruthers and A. Orridge, *J. Chem. Soc., Perkin Trans. 1*, 1977, 2411;   H. Gerlach and W. Müller, *Helv. Chim. Acta*, 1972, **55**, 2277.

28 General Strategy B: Strategy of Carbonyl Disconnections

This chapter links the carbonyl disconnections of the last 10 chapters with the general principles established in chapter 11. We shall find some new principles but the main idea is to discover why, in designing the synthesis of a particular molecule, some disconnections prove more helpful than others.

We could look at every possible C–C carbonyl disconnection and decide which we prefer. For any even moderately complex molecule, this can be an exhausting process and we shall do it for just one target molecule. Thereafter we shall choose disconnections as we go along and go back to the target only if that strategy proves poor. Pratt and Raphael[1] needed the keto-diester **1** for a synthesis of the anti-tumour compound vernolepin. Our first disconnection is easy as the α,β-unsaturated carbonyl unit suggests the classic aldol **1a** disconnection to **2**.

Compound **2** has 1,3-, 1,4-, 1,5- and 1,6-dicarbonyl relationships. Disconnecting the 1,3-diCO in the two possible directions **2a** and **2b** gives a one- or two-carbon fragment and enolates **3** and **4** that would be very difficult to control. There is in any case little simplification in either of these disconnections so we shall not pursue this strategy.

The 1,4-diCO disconnection **2c** looks promising as the required enolate **6** is of a stable 1,3-dicarbonyl compound and the electrophile is available bromoacetate **5**. Much will depend on how easy it is to make **6**.

Organic Synthesis: The Disconnection Approach. Second Edition   Stuart Warren and Paul Wyatt
© 2008 John Wiley & Sons, Ltd

The 1,5-diCO disconnection **2d** also looks promising as the required enolate **6** is again stable and the electrophile is the available enone **8**. At the moment there is little to choose between these disconnections **2c** or **2d** but the ease of making of **6** or **7** may be decisive.

Finally we can investigate the 1,6-dicarbonyl approach by reconnection **2d** to give a cyclohexene that seems destined for synthesis by a Diels-Alder reaction from isoprene **11** and the enone **10** that can probably be made by a Mannich reaction on ethyl acetoacetate.

So we have three promising approaches. But the reactions in the first two are the same: they are just done in the reverse order. So the sensible thing is to try one of those so that the starting materials can be used for the other if necessary. Pratt and Raphael found that the 1,5-diCO strategy via enolate **7** was successful. The others may be successful too. Note that the final cyclisation of **2** required only weak acid and weak base.

The Synthesis of a Lactone

Where there are structural C–X bonds in the target molecule, it makes sense to disconnect them first as we can then see the carbon skeleton displayed and count the relationships between the functional groups. So the lactone **14** has the carbon skeleton of **15** and this compound has 1,3- and 1,4-diCO relationships **15a**.

We can continue both strategies by disconnections at the branchpoint, each needing simple aryl ketones **16** and **18**, but **15b** requires a homoenolate reagent for the d^3 synthon **19** and we should rather avoid that, while **15c** needs a simple enolate **17** and we prefer that.

The keto-acid **16** is going to be made by a Friedel-Crafts reaction and the best reagent is succinic anhydride **22** so that the disconnection is outside the 1,4-diCO system as explained in chapter 25. The starting material **21** can be made by methylation of **20** with dimethyl sulfate and the enolate chosen for the last step was the organo-zinc derivative (Reformatsky reagent) of methyl bromoacetate. The methyl ester protecting group used in **23** disappears during lactonisation.[2]

Synthesis of a Symmetrical Cyclic Acetal

The keto-acetal **24** was needed for a prostaglandin synthesis.[3] Disconnection of the acetal **24a** reveals the symmetrical carbon skeleton **25** having 1,4- and 1,5-diCO relationships. There is another 1,4-diCO relationship between the two alcohols.

None of these relationships looks very promising, in part because any C–C disconnection would destroy the symmetry. We can get round this problem by using a trick that appeared first in chapter 19. We add an extra functional group (CO_2Me) to give us a 1,3-diCO relationship that can be disconnected **26** without destroying the symmetry.

This new intermediate **27** has all the 1,4- and 1,5-diCO relationships of **25** but it also has a 1,6-diCO relationship **27a** that can be reconnected to **28** again without destroying the symmetry. An adjustment of functionality gives an obvious Diels-Alder adduct **29** of butadiene and maleic anhydride **30**.

27a   1,6-diCO reconnect → 28   FGI → 29   Diels-Alder → + 30

Since the two hydrogen atoms in maleic anhydride **30** are *cis*, they must also be *cis* in the adduct **29**. Reduction and protection are needed before the oxidative cleavage so that the difference between the left- and right-hand halves of the molecule is preserved. Hydrolysis of the ester in **33** and decarboxylation in acid gave **24**.

29   1. LiAlH₄ 2. TsOH (MeO)₂CMe₂ → 31   1. KMnO₄ 2. CH₂N₂ → 32   NaH → 33

Synthesis of a Spiro Enone

Corey needed the spiro-enone **34** for his synthesis of gibberellic acid.[4] The obvious enone disconnection reveals a keto-aldehyde **35** with a 1,4-relationship between the carbonyl groups and disconnection at the branchpoint suggests some enol(ate) equivalent of the aldehyde **36** and the bromoketone **37**.

34   aldol → 35   1,4-diCO → 36 + 37

The planning for this synthesis involves the repeated disappointment of rejection of good-looking strategies. For example, the aldehyde **36** has a 1,5-diCO relationship but cannot easily be made by conjugate addition as that would require conjugate addition of an acyl anion equivalent. One alternative might be to use an allylic bromide **39** (reconnection strategy—chapter 26) and the enolate of ethyl acetoacetate **40**.

35   reconnect → 38   1,4-diCO → 39 + 40

But this strategy is doomed too as the allylic bromide will almost certainly react at its less hindered end on the vinyl group. However, we might choose the alternative branchpoint disconnection **38a** and consider conjugate addition of some vinyl-metal (copper?) derivative to the enone **41** that could be made by some aldol process from the ketone **42** and an enol(ate) of acetone **43**.

The ether **42** can obviously be made from the hydroxyketone **44** and, as we shall see in chapter 36, a good way to make 1,4-difunctionalised cyclohexanes is by reduction of a cheap aromatic compound such as quinol **45**.

This approach requires a basic kind of chemoselectivity (chapter 5) to distinguish the two phenolic hydroxyl groups in **45** so that one may be alkylated and one oxidised. Trial and error revealed that the best way was to reduce completely first and benzylate to give a mixture of unreacted diol **46**, mono-ether **48** and diether **47**. This is the statistical method (chapter 5) that is expected to give about 50% of **48** and 25% of **46** and **47**. Fortunately these can easily be separated and recycling **46** directly and **47** after debenzylation gives a good conversion. Oxidation of **48** then gives **42**. In any case, the benzylation is an early step in the synthesis and can be carried out on a large scale with such cheap materials.[5]

Corey chose a Wittig-style (HWE) reaction to control the 'aldol' process and copper-catalysed addition of vinyl Grignard for the conjugate addition. Oxidation with $NaIO_4$ and catalytic OsO_4 gave the keto-aldehyde **35** which cyclised cleanly under equilibrating conditions.

The Synthesis of Piquindone

Our final example is the heterocyclic diketone **50**, an intermediate in the Hoffmann-La Roche synthesis of piquindone **51**, an anti-psychotic agent.[6]

We might first think of removing the structural heteroatom—the ring nitrogen. With reductive amination in mind we might consider imines from **52** or amides from **53**. But these compounds have four different carbonyl groups and obviously problems of selectivity arise.

It might be better to start on the carbonyl disconnections immediately. The most obvious come from the 1,3-dicarbonyl relationship **50b, c** and suggest two keto-ester starting materials **54** and **55**.

These two intermediates **54** and **55** each have a 1,5-diCO relationship that can be disconnected in two ways. Both **54b** and **55b** disconnect a ring bond and give unsimplified starting materials **56** and **59**. But the others **54a** and **55a** achieve some simplification and suggest simple cyclic enones **57** and **60** in combination with enol(ate)s of acetone **58** or an acetate ester **61**. These are much more promising and we shall come back to them.

Incidentally, the 1,5-diCO relationship is present in the target molecule **50** too and attempts to disconnect it **50d,e** reveal a most unpromising 10-membered ring **62** or a more promising diketone **63**. Continuing the analysis of **63** would lead us back to **57** or **60**.

Returning to **57** and **60**, the α,β-unsaturated carbonyls suggest aldol-style disconnections to **64** and **65**. You may by now be reminded of something we saw in chapter 19: compounds like **64** and **65** with 1,3-diX relationships between nitrogen and a carbonyl group. There the carbonyl groups were both esters: here one must be an aldehyde and the other either an ester or a ketone.

We could presumably make **57** from **67** (compound **44** in chapter 19) by reduction and elimination. Addition of, say, acetoacetate anion should give **63**.

In fact, it isn't necessary to make **57** as, when R=Me, it is the natural alkaloid arecoline. However, when the synthesis was continued by attempted conjugate addition of the enolate of methyl acetoacetate to **57** only very low yields (12–18%) of products could be found. The problem turned out to be the base-catalysed rearrangement of **57** into the aromatic pyridone **71**.

Using **60** and malonate solved this problem. The intermediate **72** and the diketone **50** could be isolated and characterised but it was better in the manufacturing process, to continue with the formation of the drug **51** in the same process. The drug has potent and selective dopamine antagonist activity.

There remained the synthesis of **60**. An old synthesis[7] used essentially the strategy we have outlined via **65**: alkylation of $MeNH_2$ with the chloro-acetal **74**, conjugate addition of **75** to butenone **76** and cyclisation in acid solution. It was very low yielding. One reason is the poor amine synthesis by alkylation (chapter 8) and another is presumably that the acetal **77** hydrolyses to the aldehyde **65** but control in the cyclisation is poor.

The workers at Roche chose to make **60** by a completely different strategy inspired by the availability of a pyridine **78** with the same skeleton. Protection and methylation gave the pyridinium salt in very high yield and reduction and hydrolysis gave **60**. The paper[6] describing the search for a good synthesis of **51** is worth longer study.

Summary of General Approach to the Design of Syntheses

1. Convert all FGs to those based on oxygen (OH, CO etc.) by FGI or C–X disconnection so that the carbon skeleton is exposed.
2. Identify the relationships between the functional groups. This means *counting*!
3. Adjust oxidation level if necessary and disconnect using reactions from chapters 18–28.
4. Continue to examine all possible relationships (e.g. by counting both ways round a ring) until a good synthesis emerges.
5. If necessary, add extra FGs or activating groups to make reactions possible.
6. If a bad step must be included, try to make it the first step.

References

1. R. A. Pratt and R. A. Raphael, *Unpublished Observations*.
2. A. S. Dreiding and A. J. Tomasewski, *J. Am. Chem. Soc.*, 1954, **76**, 540.
3. H. Naki, Y. Arai, N. Hamanaka and M. Hayashi, *Tetrahedron Lett.*, 1979, 805.
4. E. J. Corey and J. G. Smith, *J. Am. Chem. Soc.*, 1979, **101**, 1038.
5. D. A. Prins, *Helv. Chim. Acta*, 1957, **40**, 1621.
6. D. L. Coffen, U. Hengartner, D. A. Katonak, M. E. Mulligan, D. C. Burdick, G. L. Olsen and L. J. Todaro, *J. Org. Chem.*, 1984, **49**, 5109.
7. A. Wohl and A. Prill, *Liebig's Ann. Chem.*, 1924, **440**, 139.

29 Strategy XIII: Introduction to Ring Synthesis: Saturated Heterocycles

Background Needed for this Chapter Reference to Clayden, *Organic Chemistry:* Chapter 42: Saturated Heterocycles and Stereoelectronics.

Cyclisation Reactions

This chapter is about intramolecular reactions and, in particular, about making heterocycles by cyclisation reactions. At the end of the last chapter we mentioned that the synthesis of **76** by reaction of the primary alkyl chloride **74** with $MeNH_2$ was likely to give a poor yield (numbers from chapter 28). The problem is that the product **75** is also a nucleophile and will react at a similar rate with **74** as does $MeNH_2$. The reaction is *inter*molecular and so bimolecular.

We hope you would not be surprised that the very similar reaction of **2** gives exclusively the pyrrolidine **3**. The reaction **4** is now *intra*molecular—a unimolecular cyclisation in fact—and is greatly preferred to any bimolecular processes. In fact **2** cannot be prepared as the free amine: its salts, e.g. the hydrochloride, are stable but neutralisation with base liberates **2** which promptly cyclises to **3**.

This chapter is about the advantage that cyclisations have over intermolecular reactions and therefore about the simplicity of heterocyclic synthesis. We need to look first at some details. It is not true that all cyclisations are favourable. A general cyclic amine synthesis shows that there is a large difference in the rates of formation of rings of different sizes.[1] Five-membered rings

Organic Synthesis: The Disconnection Approach. Second Edition   Stuart Warren and Paul Wyatt
© 2008 John Wiley & Sons, Ltd

are formed fastest, then six- and three-membered while four-membered rings are formed very slowly and seven-membered rings quite slowly.

k_{rel} (H$_2$O)   70   1   6 x 10^4 1 x 10^3 17

This is partly to do with the ease of the cyclisation mechanism and partly to do with the stability of the ring being formed. Three- and four-membered rings are unstable because of 'strain', that is the angles between the bonds are significantly less than the tetrahedral angle of about 109°. Five- six- and seven-membered rings are unstrained and stable with six-membered rings being the most stable as they can have a chair conformation. So cyclisation **6** to give three-membered rings is favourable because the favoured conformation is right for cyclisation and the nucleophilic N is close to the electrophilic C. Cyclisation to give four-membered rings is unfavourable because the favoured conformation **9** cannot cyclise and the conformation **10** that can cyclise has eclipsed bonds. Both cyclisations are unfavourable because the products and hence the transition states are strained.

Five-membered ring formation is very favourable as the conformation needed **4** is reasonable and transition state and product are unstrained. If you make a molecular model of a long chain and fold it round you will find that the atoms that approach each other have a 1,5-relationship. Folding a chain **12** to form a six-membered ring takes the nucleophile past the electrophile **12a** and only when the chain folds up in a chair-like fashion can cyclisation **13** occur.

So to summarise: kinetic and thermodynamic factors affect cyclisation to different extents depending on the ring size and on the reaction. Table 29.1 gives a general indication.

TABLE 29.1   Kinetic and thermodynamic factors affecting cyclisation

Ring Size	Kinetic Factors	Thermodynamic Factors
3	very favourable	unfavourable (strain)
4	unfavourable	unfavourable (strain)
5	favourable	favourable
6	moderately favourable	very favourable
7	moderately favourable	moderately favourable

Thermodynamic factors can be divided into enthalpic and entropic components. The difference between five- and six-membered rings is shown in the formation of lactones **16** and **18** from hydroxyacids **15** and **17**. Enthalpy favours the six-membered ring as the transition state is more stable but entropy favours the five-membered ring as there is a higher chance that **15** will be in a favourable conformation for cyclisation.

$\Delta H^{\ddagger}$ 80 kJ mol^{-1} $\Delta S^{\ddagger}$ –48 J deg^{-1} mol^{-1} $\Delta H^{\ddagger}$ 58 kJ mol^{-1} $\Delta S^{\ddagger}$ 108 J deg^{-1} mol^{-1}

The difference between the most kinetically favoured cyclisations is easily seen in the type of base needed to cyclise chloro-alcohols to three- and five-membered cyclic ethers (epoxides and THFs). Chloroethanol **19** cyclises only as the oxyanion: specific base is needed, i.e. a strong enough base to remove the OH proton completely. By contrast, 4-chlorobutanol cyclises by general base catalysis: the proton is removed during the cyclisation **22** and weak bases will do.[1, 2]

This means that three-membered rings can easily be made but tend to decompose under the conditions of their formation, that four-membered rings are difficult to make, that it is often difficult to stop five- and six-membered rings forming and that seven-membered rings can usually be formed when needed.

Three-Membered Rings

You are already familiar with the simple formation of epoxides **26** by the action of peroxyacids such as *m*CPBA on alkenes **27**. They can equally well be made by cyclisation of chloro-alcohols **25** as in the Cornforth addition of a Grignard reagent to an α-chloroketone and cyclisation in base.[3]

The nitrogen heterocycles, aziridines, can be made by displacement of an alcohol by an amine after activation. In their synthesis of the antitumour and antibiotic compound **30**, whose active region is the aziridine, J. P. Michael and group opened the cyclic sulfite **28** with azide ion. Reaction occurred at the allylic position and with inversion. Activation of the alcohol as a mesylate gave **29** and reduction of the azide to an amine was followed by base-catalysed cyclisation, again with inversion.[4]

28 → 29; 90% yield → 30; 60% yield

Four-Membered Rings

Though these are the least favourable cyclisations, they do happen and are usually preferred to intermolecular reactions where there is no alternative, i.e. no more favourable ring can be formed. Upjohn's analgesic and anti-depressant tazadoline **31** contains a four-membered cyclic amine, an azetidine. Simple disconnection of C–N bonds gives **32** (where X is a leaving group) and then the enone **33**, the aldol product from cyclohexanone **34** and benzaldehyde.

Rather surprisingly this strategy works.[5] It was better to use the diketone **36** (rather than **33**), made by acylation of the morpholine enamine **35** of **34** and reductive amination of with 3-aminopropanol to give **37** that is dehydrated in acid to the amine **38**. A Mitsunobu-like treatment with Ph$_3$P·Br$_2$ converts the OH to Br whereupon cyclisation of **32**; X = Br gives **31**.

They offer an even more surprising alternative in which reductive amination with benzylamine with reduction of the other ketone and cleavage of the N-benzyl bond followed by dehydration gives the simple amine **40**. Now reaction with 1,3-dibromopropane gives **31**, presumably again via **32**; X = Br.

When a four-membered heterocycle is *cis*-fused on the side of another ring, as with *syn*-**44**, which we met in chapter 12, cyclisation of the *syn*-monotosylate **42** in base is very efficient as the, usually unfavourable, conformation **10** is now the only possible one and the nucleophile and electrophile are perfectly arranged **43** for cyclisation. This observation took on a new importance when the anti-cancer compound taxol® was discovered as it also has a *cis*-fused oxetane.[6]

Five-Membered Rings

These are the most favourable of all and the precursors, such as the hydroxy acids, e.g. **15**, cannot usually be isolated, though the carboxylate salts are stable. The only important thing is to get the oxidation level of the precursor right. Using cyclic amines as examples, a fully saturated ring **45** would come from an alkylation reaction on **46**; X = a leaving group. Imines **47** or enamines **49** would come from aldehydes or ketones **48**.

Lactams **50** come from acid derivatives **51**. Compounds such as amino esters **53** are not stable as the free amine but are usually isolated as salts such as the hydrochloride **52**. When treated with base, **52** gives the free amine **53** which promptly cyclises to the lactam **50**.

For examples, we shall include sulfur heterocycles. The unsaturated ring **54** does have a double bond but it is not next to the heteroatom: it is an allylic rather than a vinylic sulfide. So two disconnections at the alcohol oxidation level suggest the doubly allylic starting material **55**. As it happens, we made the diol **56** in chapter 16 so we simply have to turn the OH groups into leaving groups and cyclisation with Na_2S gives the heterocycle.[7]

When Metzner[8] wanted to make the C_2 symmetric sulfide **57** he looked no further than the diol **58** which was available as a single enantiomer by asymmetric reduction. Conversion into a *bis*-mesylate **59** (not isolated) and double displacement with Na_2S gave the thiolane **57**. The first displacement must be *inter*molecular but the second is *intra*molecular so it is much faster than reactions with other molecules of **59**. Both displacements go stereospecifically with inversion.

A combination of C–C and C–X disconnections can lead to a short synthesis. The sulfide **60** was needed by Woodward as an intermediate in a synthesis of biotin.[9] Immediate C–S disconnections lead to an unlikely and very reactive compound **61**. If instead the 1,3-diCO disconnection **60a** is done first, the same C–S disconnections can be done to give simple starting materials.

Woodward chose to do the conjugate addition second because thiolacetic ester **63** was available. The diester **62** cyclised under equilibrating conditions to give **60**.

A compound with two heteroatoms in the ring **66**, needed for a synthesis of mannicone **65**, illustrates the other two oxidation states. Disconnecting both C–N bonds noting that C-1 is at the acid oxidation level while C-3 is an enamine and so at the ketone oxidation level gives the keto-ester **67** and hydrazine **68**. Where two heteroatoms are joined in a ring, it is usually better not to disconnect the bond between them but to look for a starting material containing both.

Now the 1,3-diCO relationship can be disconnected in two ways. One **67b** would require a condensation between two esters that is difficult to control but the other **67a** will be regioselective for reasons we explained in chapter 20.

The synthesis[10] is straightforward: the condensation is indeed regioselective and it doesn't matter which way round hydrazine reacts or indeed which carbonyl compound reacts first. The enamine will form so as to be in conjugation with the remaining carbonyl group.

Six-Membered Rings

The same methods work well with six-membered rings so we shall look at a few instructive examples. A dramatic demonstration of the advantages of intramolecular reactions comes in the synthesis of tetramethyl piperidone **72**. Removal of the nitrogen with conjugate addition of ammonia to the dienone **73** opens the possibility of a double aldol disconnection to reveal three molecules of acetone.

Treatment of acetone with ammonia and the mild dehydrating agent calcium chloride at room temperature gives the piperidone **72** in one step.[11] Presumably the acetone dimerises and trimerises but the incipient polymerisation is nipped in the bud by the capture of one or more of these intermediates by ammonia and the formation of the only possible stable six-membered ring heterocycle **72**. The yield (48%) seems poor but excess acetone is recovered in the isolation and the yield is 70% if that is taken into account. In any case, making 430 g of **72** from 1 kg acetone is a cheap process with so much 'added value'.

The morpholine derivative **75** has an obvious amide disconnection to **76** and a less obvious 1,2-diX disconnection at the ether to **77**. This is obviously an epoxide **78** adduct with ammonia.

But supposing things are not so simple. The compound **79**, with a substituent on the other carbon atom and required as a single enantiomer, starts the same way but when we get to **81** we cannot go further using an epoxide.

Fortunately **81** is phenylalaninol—the alcohol from the amino acid phenylalanine. The workers decided to acylate at nitrogen first and so used chloracetyl chloride to make **77** which cyclised on treatment with base.

This compound **79** was needed to make a coenzyme analogue **86**. Alkylation on oxygen by Meerwein's reagent gave the activated imino-ether **84** which reacted with phenylhydrazine to give the hydrazone salt **85**. Now insertion of a single carbon atom as an orthoester HC(OMe)$_3$ gave the triazolium salt **86**. Note that the two bonded nitrogen atoms are added at once and that addition of a single carbon atom in a cyclisation reaction works superbly well.[12]

With two heteroatoms in the six-membered ring, it is again useful to identify a compound containing both. The nucleic acid base uracil **87** contains a molecule of urea **88** in its structure but removing that leaves an awkward synthon **89**. The question is; how are we to add to the enone and get the double bond back? The solution is to use the acetylenic acid **90**. Heating **90** and **88** together in acid gives uracil in 65% yield.[13]

Seven-Membered Rings

Seven-membered nitrogen-containing heterocycles, especially those fused to benzene rings, are important in drug design. The famous tranquillisers Librium® **91** and Valium® are based on such a ring system with two nitrogens **91**. In the search for similar compounds,[14] workers at Hoffmann-La Roche chose analogue **92**. Disconnection of two of the ring C–N bonds (it doesn't alter things if you disconnect more) reveals the oxime **93** of an aromatic ketone and chloroacetyl chloride **94**.

In practice the oxime **96** of the amino ketone **95** can be used in the cyclisation reaction to make an intermediate **97** that can be alkylated with various alkylating agents to give **92**. The more nucleophilic amine is acylated and cyclisation to the less nucleophilic oxime nitrogen gives **97**.

Compounds such as **98** with only one nitrogen in the ring are more interesting synthetically and are needed for an anti-HIV drug.[14] Initial C=C disconnection is followed by a C–N disconnection between the ring and the nitrogen **99**. This is possible because nucleophilic aromatic substitution (S_NAr) works well on aryl fluorides with *ortho* or *para* electron-withdrawing groups[15] such as the aldehyde in **100**.

We have already explained that compounds like **101** are unstable and cyclise rapidly to the lactam. So the workers at Takeda used the lactam **102** as the starting material. Opening the lactam with NaOH gave the anion **103** of **101** that added to **100** to give **99** and hence that aldol product **98** in base. Cyclisation to a seven-membered ring is preferred to intermolecular reactions.

The related benzazapinones **105** can be made by formation of C–N bonds in two different ways.[16] Amide formation from compounds like **104** is not surprising but reductive amination between the amide nitrogen and the aldehyde in **106** is a testament to the efficiency of cyclisations even when a seven-membered ring is the product.

When workers at GlaxoSmithKline wanted to make **107** as an intermediate in the synthesis of a drug for the treatment of osteoporosis, they chose the double disconnection **107** because they already had a way of making single enantiomers of the diacid **108**.

Imine formation of the diester **110** with the amine **109** and Lewis acid catalysis gave **111** and completion of the reductive amination with NaBH(OAc)$_3$ gave the free amine that cyclised on refluxing in toluene. Though the cyclisation required these relatively vigorous conditions, it still demonstrates the preference for seven-membered ring synthesis over intermolecular reactions or, in this case, an eight-membered ring.

References

1. A. J. Kirby, *Adv. Phys. Org. Chem.*, 1980, **17**, 183.
2. B. Capon, *Quart Rev.*, 1964, **18** 64.
3. R. A. Barnes and W. M. Budde, *J. Am. Chem. Soc.*, 1946, **68**, 2339.
4. J. P. Michael, C. B. de Koning, R. L. Petersen and T. V. Stanbury, *Tetrahedron Lett.*, 2001, **42**, 7513.
5. J. Smuszkovicz, *Eur. Pat.*, 85,811; *Chem. Abstr.*, 1984, **100**, 6311.
6. G. Kinast and L.-F. Tietze, *Chem. Ber.*, 1976, **109**, 3626.
7. R. H. Everhardus, R. Gräfing and L. Brandsma, *Recl. Trav. Chim. Pays-Bas*, 1976, **95**, 153; B. A. Trofimov, S. V. Amosova, G. K. Musorin and M. G. Voronkov, *Zh. Org. Khim.*, 1978, **14**, 667; *Chem. Abstr.*, 1978, **88**, 190507.
8. K. Julienne, P. Metzner, V. Henryon and A. Greiner, *J. Org. Chem.*, 1998, **63**, 4532.
9. R. B. Woodward and R. H. Eastman, *J. Am. Chem. Soc.*, 1946, **68**, 2229.
10. L. H. Sternbach and E. Reeder, *J. Org. Chem.*, 1961, **26**, 4936.
11. G. Sosnovsky and M. Konieczny, *Synthesis*, 1976, 735.

12. R. L. Knight and F. J. Leeper, *J. Chem. Soc., Perkin Trans. 1*, 1998, 1891.

13. R. J. DePasquale, *J. Org. Chem.*, 1977, **42**, 2185.

14. T. Ikemoto, T. Ito, A. Nishiguchi, S. Miura and K. Tomimatsu, *Org. Process Res. Dev.*, 2005, **9**, 168.

15. Clayden, *Organic Chemistry*, chapter 23.

16. M. D. Wallace, M. A. McGuire, M. S. Yu, L. Goldfinger, L. Liu, W. Dai and S. Shilcrat, *Org. Process Res. Dev.*, 2004, **8**, 738.

30 Three-Membered Rings

Background Needed for this Chapter Reference to Clayden, *Organic Chemistry:* Chapter 40: Synthesis and Reactions of Carbenes.

Chapters 30–37 are concerned with the synthesis of carbocyclic rings. The disconnections are therefore of C–C rather than C–X bonds and the choice is correspondingly greater. We start in this chapter with three-membered rings and work our way upwards to six-membered. But the principles, particularly of cyclisation, remain the same as in chapter 29.

Cyclopropanes by Alkylation of Enolates

Three-membered rings are kinetically favoured but thermodynamically unstable so that they are often destroyed under the conditions of their formation. Since most carbonyl condensations are reversible, they are generally not good routes to three-membered rings. But the alkylation of enol(ate)s is usually irreversible so that these can be excellent methods.

Cyclopropyl ketones **1** can be made by cyclisation of some derivative of the γ-hydroxy-ketone **2**. Notice that we proposing to make a three-membered *carbo*cyclic ring from an easily made three-membered *hetero*cyclic ring.

Since we have already made compounds like **2** in chapter 25 from β-keto-esters, it makes sense to use the same strategy here. Addition of ethylene oxide **3** to the enolate of **5** gives the lactone **6** directly and treatment with HBr accomplishes decarboxylation and formation of the bromide **7** in one pot.[1] Vogel[2] uses the chloroketone to make **1**; R=H in 82% yield by this method with NaOH for the base.

Organic Synthesis: The Disconnection Approach. Second Edition   Stuart Warren and Paul Wyatt
© 2008 John Wiley & Sons, Ltd

A more dramatic example is the synthesis of *cis*-chrysanthemic acid **11**, the basis of most modern insecticides, from dimedone **8**, whose synthesis we discussed in chapter 21. Methylation between the two carbonyl groups gives **9**, with the complete skeleton of **11**—a little reorganisation of the atoms is needed. Treatment with bromine and base gives the inevitably *cis*-fused bicyclic dione **10** and a further three simple steps produce chrysanthemic acid.[3]

Some explanation is needed! Treatment of **9** with base and bromine must produce the potassium enolate of the bromoketone that cyclises **12** to form the three-membered ring. Reduction presumably gives the *exo*—alcohol[4] **13** whose tosylate can fragment with hydroxide **14** to give **11**.

A much less promising cyclisation gives the biologically patterned insecticide permethrin[5] **17**. The enolate of the ester in the starting material **15** must cyclise by displacement of chloride at a tertiary centre. Cyclisation to form three-membered rings can be remarkably favourable.

The simple bicyclic amines **18** are drug candidates with Merck for treatment of pain.[6] Disconnection to one of the amino alcohols **19** suggests that the anion of the nitrile **23** might be used to make two C–C bonds in the cyclopropane by alkylation of epichlorhydrin **22** both at the epoxide and at the chloride. If this sequence works, it can give only the stereochemistry required.

The workers at Merck were able to go from start to finish in a single vessel. Base treatment of **22** + **23** led to attack at the epoxide end of **22** to give the anion **24**. This is in equilibrium with the carbanion stabilised by Ar and CN and cyclises to an epoxide that gives a 15:85 mixture of the diastereoisomers **25** and **20**. Reduction with borane gave a mixture of the amino alcohols which was converted into a mixture of the chlorides **26** as their hydrochloride salts. Raising the pH to >8.5 gave enough of the free base of **26** to cyclise to **18**. The two drugs bicifadine (Ar = *p*-tolyl) and DOV21947 (Ar = 3,4-dichlorophenyl) were formed in acceptable overall yield for this multi-step operation. The *trans* compound obviously cannot cyclise and was removed by crystallisation.

Carbene Insertion into Alkenes

The cyclisation methods we have used so far all depend on the simple disconnection **27a** into a carbonyl group and an alkylating agent in the same molecule **28**. But the same general class of molecule **27** can be made a very different way that is revealed by disconnection of two C–C bonds **27b** to suggest an alkene **29** and a carbene **30**.

Carbenes have divalent carbon with a lone pair and hence only six electrons in the outer shell of the carbon atom. They are normally electrophilic and can form two bonds at once with a π-system.[7] One way to make carbenes is by loss of nitrogen from diazocompounds such as diazoketones **33**. The formation of very stable nitrogen is initiated by heat or light and compensates for the formation of the unstable carbene **30**. Diazoketones are easily made by acylation of diazomethane with an acid chloride **31**. Loss of a very acidic proton from the diazonium salt **32** gives **33**. Normally the diazoketone and the alkene are combined and treated with heat or light.[8]

This will obviously be easier if both reactive centres are in the same molecule so disconnection of the tricyclic ketone **34** reveals a diazoketone **35** that can be made from the acid **36**. The branchpoint disconnection would require a d^3 reagent.

34   35   36   37

This strategy did not appeal to Ruppert and White[9] who preferred to make **36** by a chain extension route. A Reformatsky reagent gave **39** and dehydration and alkylation of malonate gave **36**. Treatment with oxalyl chloride $(COCl)_2$ followed by reaction with diazomethane gave **35** which duly cyclised to **34** with copper catalysis.

38   39   40   41

Carbenes by α-Elimination

You will be familiar with β-elimination that leads to alkenes (chapter 15) but α-elimination gives carbenes. The simplest example is dichlorocarbene **44** made by treatment of chloroform **42** with base. The carbanion **43** decomposes by loss of chloride ion to give the neutral but unstable carbene **44** that should be released in the presence of the alkene. So cyclohexene **45** gives the dichloro-cyclopropane **46** using NaOH as the base and an ammonium salt as a phase-transfer catalyst.[10]

42   43   44   45   46; 62% yield

A more interesting example is the aryl-cyclopropane **47**. Removing the carbene we must put back the HCl lost in the α-elimination to reveal **48**. An example is with the benzylic chloride **49** that gives the bicyclic compound **47** in good yield and with high diastereoselectivity (10:1) in favour of the *endo* isomer shown. The strong and very hindered base **50** is used.[11]

47   48   49   50   47; 82% yield

Metal Complexes of Carbenes

Some metals form stable complexes of carbenes, a notable example being rhodium. Treatment of the diazocompound with catalytic rhodium acetate gives the cyclopropane **52** via the carbene complex **53**. Notice that the carbene inserts in only the less hindered of the two alkenes.[12] This style of chemistry is treated in more detail in *Strategy and Control*.

51   cat [Rh(OAc)$_2$]$_2$ **52; 71% yield**   **53**

Metal Carbenoids

Related to metal complexes are metal carbenoids such as those formed when zinc reacts with di-iodomethane. In early examples, such as the efficient cyclopropanation of cyclohexenone **54**, the zinc was activated by some copper.[13] The active reagent is the zinc σ-complex **56**. One might suppose an α-elimination **57** would occur to give the carbene, but this is apparently not so. The active reagent **56**, nearly but not quite a carbene, is known as a *carbenoid*.

54   CH$_2$I$_2$ / Zn/Cu   **55; 90% yield**   Zn   **56**   **57**   CH$_2$ + ZnI$_2$

A better version involves diethyl zinc as the source of the metal and again, simple alkenes such as **58** can be converted into cyclopropanes.[14] The active reagent is probably **60**.

58   CH$_2$I$_2$ / Et$_2$Zn   **59; 86% yield**   Et$_2$Zn   **60**

However, this method really hit the headlines when it was used on allylic alcohols **61** and became known as the Simmons–Smith reaction.[15] If there is stereochemistry at the alcohol **63**, the cyclopropane is formed on the same side as the OH group **64** suggesting that the alcohol guides the zinc carbenoid into the alkene.

61   CH$_2$I$_2$ / Zn/Cu   **62**   **63**   CH$_2$I$_2$ / Zn/Cu   **64**

In their synthesis of the anti-leukaemia compound steganone, Magnus, Schultz and Gallagher used the allylic alcohol **65** to make the cyclopropane **66**. The extra carbon atom was incorporated into the eight-membered ring of steganone.[16]

65   **66; 74% yield**

An example with stereochemistry comes in the synthesis of halicholactone by Takemoto.[17] The diene **67** (the various R groups are protecting groups) gives only one cyclopropane: the allylic alcohol alone reacts and the cyclopropane appears on the same face of the alkene as the allylic OH group **68**.

67   **68; 69% yield**

All these methods using carbenes, metal complexes of carbenes or carbenoids are stereo-specific in that the geometry of the alkene is faithfully reproduced in the stereochemistry of the cyclopropane so *trans*-**67** gives *trans*-**68** specifically. They can also be stereoselective, particularly the Simmons–Smith on allylic alcohols: thus the cyclopropane in **68** is on the same side of the alkene as the OH group in **67**. We now come to a widely used method that is not stereospecific on the alkene.

Sulfonium Ylid Chemistry

The simplest sulfur ylids are formed from sulfonium salts **69** by deprotonation in base. These ylids react with carbonyl compounds to give epoxides.[18] Nucleophilic attack on the carbonyl group **70** is followed by elimination **71** of dimethylsulfide **72** and formation of the epoxide **73**. You should compare diagram **71** with diagram **23** in chapter 15. The phosphonium ylid reacted by formation of a P–O bond and an alkene in the Wittig reaction. The sulfonium compound reacts by formation of a C–O bond **71** as the S–O bond is much weaker than the P–O bond. The sulfonium salt **69** can be reformed by reaction of **72** with MeI.

69   **70**   **71**   **72**   **73**

So what has this got to do with cyclopropanes? If sulfur ylids react with enones either the epoxide **74** or the cyclopropane **76** may be formed.[19] The general rule is that sulfonium ylids from **69** give epoxides but sulfoxonium ylids give cyclopropanes **76**.

74   **75**   **76**

The sulfoxonium ylid **78** is more stable and is therefore liable to do conjugate rather than direct addition (chapter 21). The intermediate eliminates dimethyl sulfoxide **79** to give the cyclopropane **76**. The intermediate is long lived and the single bond that was the alkene can rotate so the geometry of the alkene is lost. In this case we expect the more stable *trans* cyclopropane to be formed by choice.

These reactions may show considerable selectivity. Corey and Chaykovsky[19] give an example with the terpene carvone **80**. The ylid **78** is made with NaH and reacts only with the enone and not with the unconjugated alkene. The product is one diastereoisomer **81** as the ylid has added to the opposite side of the ring to the only substituent. It also has retained the stereochemistry of the *cis* alkene but that is inevitable as 3/6 ring fusion must be *cis*.

A more complex example comes in the synthesis of halicholactone by Wills.[20] A diene **82** is again used and again only the conjugated alkene gives a cyclopropane **83**. The reaction stereoselectively gives a 5:2 ratio of **83** and the other *trans* cyclopropane with the ring below the chain. Notice that the allylic alcohol is blocked so there is no Simmons–Smith style direction. The geometry of the alkene again appears to be retained but that is by choice. The product is cyclised to give the nine-membered lactone **84** where R is the cyclopropane-containing side chain.

In all these carbene-related methods the disconnection is the same (as in **34** or **47**) with a choice over which pair of bonds in the three-membered ring you prefer to disconnect. Or you could think which alkene you would rather make. In most cases we prefer to remove a CH_2 group if possible as the reagents CH_2I_2 or **78** are much easier to come by. However the diazo **34** and simple carbene **47** examples show that this is not a rule.

References

1. J. M. Conia, *Angew. Chem. Int. Ed.*, 1968, **7**, 570.

2. *Vogel*, page 1090.

3. A. Krief, D. Surleraux and H. Frauenrath, *Tetrahedron Lett.*, 1988, **29**, 6157; A Krief, G. Lorvelec and S. Jeanmart, *Tetrahedron Lett.*, 2000, **41**, 3871.

4. Clayden, *Organic Chemistry*, chapter 33.

5. M. Elliott, A. W. Farnham, N. F. Janes, P. H. Needham, D. A. Pulman and J. H. Stevenson, *Nature (London)*, 1973, **246**, 169; P. D. Klemmensen, H. Kolind-Andersen, H. B. Madsen and A. Svendsen, *J. Org. Chem.*, 1979, **44**, 416.

6. F. Xu, J. A. Murry, B. Simmons, E. Corley, K. Finch, S. Karady and D. Tschaen, *Org. Lett.*, 2006, **8**, 3885.

7. Clayden, *Organic Chemistry*, chapter 40.

8. S. D. Burke and P. A. Grieco, *Org. React.*, 1979, **26**, 261.

9. J. F. Ruppert and J. D. White, *J. Chem. Soc., Chem. Commun.*, 1976, 976.

10. *Vogel*, page 1110.

11. R. A. Olofson and C. H. Dougherty, *J. Am. Chem. Soc.*, 1973, **95**, 581.

12. C. Meyers and E. M. Carreira, *Angew. Chem. Int. Ed.*, 2003, **42**, 694.

13. J.-C. Limasset, P. Amice and J. M. Conia, *Bull. Soc. Chim. Fr.*, 1969, 3961.

14. K.-J. Stahl, W. Hertzsch and H. Musso, *Liebig's Ann. Chem.*, 1985, 1474.

15. H. E. Simmons, T. L. Cairns, S. A. Vladuchick and C. M. Hoiness, *Org. React.*, 1973, **20**, 1; A. B. Charette and A. Beauchemin, *Org. React.*, 2001, **58**, 1.

16. P. Magnus, J. Schultz and T. Gallagher, *J. Am. Chem. Soc.*, 1985, **107**, 4984.

17. Y. Takemoto, Y. Baba, G. Saha, S. Nakao, C. Iwata, T. Tanaka and T. Ibuka, *Tetrahedron Lett.*, 2000, **41**, 3653.

18. P. Helquist in *Comprehensive Organic Chemistry*, **4**, 951.

19. E. J. Corey and M. Chaykovsky, *J. Am. Chem. Soc.*, 1965, **87**, 1353.

20. D. J. Critcher, S. Connolly, M. F. Mahon and M. Wills, *J. Chem. Soc., Chem. Commun.*, 1995, 139.

31 Strategy XIV: Rearrangements in Synthesis

Background Needed for this Chapter Reference to Clayden, *Organic Chemistry:* Chapter 37: Rearrangements.

If the carbon framework of a TM is difficult to construct, one strategy is to make a slightly different framework by conventional reactions and rearrange it into the framework we want. Methods involving rearrangement range from simple chain extensions to deep-seated skeletal rearrangements very difficult to analyse.

Diazoalkanes

In the last chapter we met diazoalkanes as sources of carbenes in the synthesis of three-membered rings. These same diazoalkanes are useful reagents for rearrangements via carbenes or carbocations.

Chain Extension with Diazoalkanes: The Arndt–Eistert Procedure

If an acyldiazoalkane **2**, made from a carboxylic acid, is treated with heat, light or a metal to give a carbene in the absence of a carbene acceptor, the carbene rearranges **3** by migration of the side chain (R) to give a ketene **4**, a strong electrophile that gives derivatives of the homologous carboxylic acid such as the ester **5**. The chain length has been increased by one carbon atom. Ketenes are discussed more fully in chapter 33.

This extension, known as the Arndt–Eistert procedure,[1] is useful if the relationship between functional groups is unhelpful in the TM but becomes helpful if the chain is retrosynthetically shortened. Other methods, such as cyanide displacement, also increase the chain length by one carbon and we saw a chain extension by two carbon atoms in the last chapter. The disconnections are strange: both C–C bonds between R and CO are made in the reaction so we must disconnect both **5a**. You might like to think of this as a reconnection strategy (chapter 26) or as an extrusion of a CH_2 group.

Organic Synthesis: The Disconnection Approach. Second Edition   Stuart Warren and Paul Wyatt
© 2008 John Wiley & Sons, Ltd

In chapter 27 we analysed the synthesis of the bicyclic lactone **7** needed by Eschenmoser for his vitamin B_{12} synthesis.[2] The next step was a chain extension by the Arndt–Eistert procedure to the ester **9**.

The unsaturated ester **10** would be made by dehydration of the tertiary alcohol **11**. But this has an unhelpful 1,4-diCO relationship that leads to a homoenolate **13** that we should rather avoid.

Chain extension **11a** solves the problem. The homoenolate **13** becomes an enolate **15** with many possibilities.

Smith[3] used a Reformatsky reagent as the enolate equivalent and everything went according to plan, especially the yield on the chain extension step.

Diazoalkanes in Ring Expansion and Contraction

Related to simple chain extension is ring expansion and contraction useful because some ring sizes are easier to make than others. So available cyclohexanone can be expanded into cycloheptanones such as the useful keto-ester **20** with an activated position for enolate reactions. The reagent is ethyl diazoacetate **18** readily available from glycine esters. Addition to the ketone **18** automatically produces an oxyanion and a diazonium leaving group that collaborate **19** in the migration.[4] The disconnection is again extrusion of a carbon atom.

Disconnection of the bicyclic diketone **21** starts reasonably to reveal **22** but the two carbonyl groups are now 1,6-related suggesting a reconnection strategy (chapter 27). But this is impossible as the bridgehead alkene **23** is too strained to exist. However, if we extrude the carbon atom in the seven-membered ring between the ketone and the branchpoint **22a**, we get a new ketoester **24** with a 1,5 relationship that can be made by conjugate addition (chapter 21).

An enamine is ideal for the conjugate addition. We could have used diazomethane for the ring expansion but a better idea is to make it intramolecular by using a diazoketone **28** easily made from the free acid **27** with diazomethane. This time the rearrangement was initiated with Meerwein's salt and only the more substituted carbon atom migrates.[4]

For ring contraction, the diazo group has to be on the cyclic ketone and the reaction resembles the ketene formation at the start of this chapter. The four-membered ring in the natural product junionone **29** is difficult to make but Wittig disconnection suggests a simpler aldehyde **30** and, if we could somehow put the carbonyl group back into the ring using a diazoketone **31**, we could start with a simple cyclopentanone **32**.

The diazocompound was made with tosyl azide and it proved better to add an activating group (CHO in its enol form in **33**) for this. Rearrangement in light gave a ketene that picked up methanol to give the ester **34** and the rest is straightforward.[5] Ring contraction of five- to four-membered rings is unusual because of the increased strain but the gain in forming nitrogen compensates.

The Pinacol Rearrangement

In chapter 24 we saw that carbonyl compounds dimerise by a radical reaction when electrons are transferred to them from metals. The typical 'pinacol' **37** formed from acetone **36** is important because it rearranges[6] in acid to give a tertiary alkyl ketone **38** known as 'pinacolone'. The key step is a methyl migration as one of the OH groups is lost **39**.

Though restricted by the need for symmetry, this is a useful approach to *t*-alkyl ketones which are otherwise difficult to make.[7] The crowded alkenes **40** can be made by dehydration of alcohols **41** and hence from the ketone **42** and RLi or RMgX. As **42** has a *t*-alkyl substituent it is a candidate for the pinacol approach.

The easiest way to do the disconnections is to reverse the rearrangement and there are two ways to do this **42a** and **42b**. Diol **44** can be made by pinacol dimerisation of cyclopentanone **43** while diol **45** would be the product of dihydroxylation of the alkene **46**.

In fact Corey[8] chose strategy **a** using the pinacol dimerisation to make the diol **44** and the pinacol rearrangement to make the spirocyclic ketone **42**.

The product **40** was used in a synthesis of the Columbian frog venom perhydro-histrionicotoxin **49** and that also used the Beckmann rearrangement of the oxime **47**. Note that only the group *anti* to the N–O bond migrates and that it does so with retention. You can see that each ring was expanded in this synthesis.

Rearrangements of Epoxides

The limitation of the pinacol is the need for symmetry. This section and the next suggest ways of avoiding this problem. Unsymmetrical epoxides are easily made from alkenes and open with Lewis acid catalysis to give the more substituted of the two possible cations.[9] Even such a weak Lewis acid as LiBr opens the epoxide **51** to give the tertiary cation **52** which rearranges by ring contraction to the aldehyde **53**. The authors prefer to have the bromocompound **54** as an intermediate.[10]

A more exciting example is the epoxide **56** of natural α-pinene **55** that rearranges to the unsaturated aldehyde **57** in excellent yield.[11] Epoxide opening to give the more substituted carbocation is followed by rearrangement **59** and then fragmentation **58**. Note that the expansion of the strained four-membered ring is preferred to any alternatives.

Semi-Pinacol Rearrangements

One limitation remains that would apply to pinacol rearrangements and unsymmetrical 1,2-diols: only the more highly substituted cation would be formed. We can get round that by selectively functionalising the less substituted alcohol with a sulfonate leaving group. Pinacol rearrangement of the diol **61** would lead to migration of H or Me to the tertiary centre. But if the secondary alcohol is selectively mesylated **62**, rearrangement gives the ketone **63** by migration of an R group **64**. This sequence was carried out with single enantiomers as the starting material is natural lactic acid.[12] The disconnections for epoxide and semi-pinacol rearrangements are the same as for the pinacol: just reverse the rearrangement. But it is not straightforward to see that this solution is available.

The Favorskii Rearrangement

The rearrangements we have seen so far are all essentially cationic even though no cationic intermediate may be involved. By contrast the Favorskii rearrangement is anionic—almost every intermediate is an anion. Halogenation of cyclohexanone gives the α-chloroketone **66**. Treatment of such compounds with nucleophilic alkoxide gives ring-contracted esters **67**. The enolate of **66** cyclises **69** to give an unstable cyclopropanone that reacts immediately with the alkoxide to cleave one of the weak C–C bonds in the three-membered ring.

Cyclopropanone cleavage with elimination **72** can also lead to ring contraction as in the synthesis of the *trans* acid **74** from natural pulegone[13] **70**. Bromination gives the unstable dibromide **71** that is immediately treated with ethoxide to initiate the Favorskii rearrangement. The product is a mixture of *cis* and *trans* isomers of the ester **73** but hydrolysis under vigorous conditions (reflux in aqueous ethanol) epimerises the ester centre and gives exclusively the *trans* acid **74**.

Once again the only reasonable way to 'disconnect' such an ester is to reverse the rearrangement in your mind, adding the halide at a reasonable position. So **67** might be made by reversing the Favorskii rearrangement **75** and only when you see the simple starting material **66** can you appreciate that this looks a good strategy. Another way would be to draw the cyclopropanone **76** from which **67** could be made. But rearrangements are difficult to visualise retrosynthetically.

References

1. W. E. Bachman and W. S. Struve, *Org. React.*, 1942, **1**, 38.

2. A. Eschenmoser and C. E. Winter, *Science*, 1977, **196**, 1410.

3. A. B. Smith, *J. Chem. Soc., Chem. Commun.*, 1974, 695.

4. W. L. Mock and M. E. Hartman, *J. Am. Chem. Soc.*, 1970, **92**, 5767.

5. A. Ghosh, U. K. Bannerjee and R. V. Venkateswaran, *Tetrahedron*, 1990, **46**, 3077.

6. *Vogel* pages 527 and 623.

7. A. P. Krapcho, *Synthesis*, 1976, 425;   B. Rickborn, *Comp. Org. Synth.*, **3**, 721.

8. E. J. Corey, J. F. Arnett and G. N. Widiger, *J. Am. Chem. Soc.*, 1975, **97**, 430.

9. B. Rickborn, *Comp. Org. Synth.*, **3**, 733.

10. B. Rickborn and R. M. Gerkin, *J. Am. Chem. Soc.*, 1971, **93**, 1693.

11. J. B. Lewis and G. W. Hendrick, *J. Org. Chem.*, 1965, **30**, 4271.

12. K. Suzuki, E. Katayama and G. Tsuchihashi, *Tetrahedron Lett.*, 1983, **24**, 4997;   G. Tsuchihashi, K. Tomooka and K. Suzuki, *Tetrahedron Lett.*, 1984, **25**, 4253.

13. *Vogel*, page 1113.

32 Four-Membered Rings: Photochemistry in Synthesis

Background Needed for this Chapter Reference to Clayden, *Organic Chemistry:* Chapter 35: Pericyclic reactions I: Cycloadditions.

In chapter 29 we concluded that four-membered rings are uniquely difficult to make: they are strained with ring angles of about $90°$ and the most favourable conformation of the starting material cannot cyclise. Four-membered rings can occasionally be formed by ordinary cyclisations. The double alkylation of malonate **1** with dibromopropane gives the cyclobutane **2**. But Perkin found in his original work on carbocyclic rings[1] that double alkylation of acetoacetate **3** was successful for all ring sizes from three to seven except four. The enol ether **4** was formed instead of a cyclobutane. It is easy to see how the enolate of the intermediate **5** is ideally arranged to form **4** but not to form a cyclobutane.

Photochemical Cycloadditions

So special reactions are often used to make cyclobutanes. In the next chapter we shall see that thermal cycloadditions of alkenes with ketenes give four-membered rings, but the commonest method is photochemical cycloaddition. You are already aware that Diels-Alder reactions (chapter 17) occur easily when a diene **6** and a dienophile **7** are heated together and six-membered rings **8** are formed. Have you ever wondered why four-membered rings **9** are not formed instead? Orbital symmetry allows cycloadditions involving six π-electrons but not those involving four π-electrons.[2]

The 2 + 2 cycloadditions do occur in the excited state so these are photochemical reactions.[3] They work best if one component (usually an enone) absorbs the light to form the excited state

Organic Synthesis: The Disconnection Approach. Second Edition   Stuart Warren and Paul Wyatt
© 2008 John Wiley & Sons, Ltd

while the other (usually a simple alkene) reacts in its ground state. Even ethylene reacts with conjugated enones such as **10** under irradiation[4] to give reasonable yields of the cyclobutane **11**. The stereochemistry of H and Me at the ring junction is determined partly by the fact that they are already *cis* in the starting material **10a** and partly by the difficulty of making *trans* 4/6 fused system.[5]

Both components may be functionalised so disconnection of the middle ring of **12** leads to the greatest simplification suggesting two simple starting materials **13** and **14**. Irradiation of the mixture does indeed give[6] a good yield (70%) of **12**. The stereochemistry of the B/C ring junction must be *cis* **12a**, as two four-membered rings must be *cis* fused, but that of rings A/B follows the guidelines of **11**. The relative stereochemistry of the two *cis* junctions, i.e. that of rings A and C, is chosen to give least steric hindrance. There is no *endo* rule. Because both compounds have the alkene conjugated to a carbonyl, the minor product is the dimer of **14**.

Most cyclobutanes offer a choice between two 2 + 2 disconnections and the choice can often be made by considering the availability of the starting materials. We already know from **11** that compounds like this can be made by irradiation of ethylene with enones, here **15**, so we shall focus on the alternative **16b**. The starting material for an intramolecular photo-cycloaddition would be dienone **17**. It was decided to make this by oxidation of **18** because this alcohol was easy to make.[7]

The idea was to add vinyl Grignard **19** to the aldehyde **20** which could be made by allylation of isobutyraldehyde **21**. Oxidation to the ketone might be carried out either before or after the cycloaddition.

In fact, the allylation was carried out by the Claisen rearrangement (chapter 35) and the cycloaddition on the alcohol **18** catalysed by Cu(I). The product was a mixture of the major isomer **23** with some of the *exo*-alcohol (OH up as drawn). This is irrelevant as both alcohols oxidise to the ketone **16**. The stereochemistry at the ring junction can only be *cis*.

Regioselectivity

Only one regioselectivity is possible in the cyclisation of **18** but in intermolecular reactions we need to consider which way round it will go. So the unsymmetrical alkene **24** could add to the very unsymmetrical enone **25** to give **26** or **27**. In fact the reaction gives[8] virtually 100% of **27** without a trace of **26**.

Some big steric or electronic factor is clearly at work. Though the alkene **24** is hindered at one end, the enone barely is and it is electronic factors that dominate. The natural polarity of the alkene is to be nucleophilic at the CH_2 group **24a**. So in the thermal reaction (that doesn't happen) it could attack the electrophilic end of the enone **25a**. One way to predict photochemical 2 + 2 cycloadditions is to suppose that the excited state of the enone reverses the natural polarity from **25b** to **25c** and the new electrophilic end now combines with the alkene **24**. As the alkene is not excited, it behaves in the normal way **24a**. This is of course a simplification but it works.[9]

Intramolecular reactions occur by the same principle if they can, but the contortion required to get the bridged cyclobutane **28** from **29** is too much and the molecule prefers to add the 'wrong' way round and give the fused structure **30** instead.[10]

But amazing contortions are possible. Photocycloaddition of the allene **31** unites just one of the allene bonds with the conjugated alkene to give the very strained cyclobutane **32**. Diagram

31a should make it clear how this distortion gives **32**. This product **32** was perfect for Hiemstra's synthesis of solanoeclepin A **33**, an extraordinary compound that is exuded from growing potatoes and causes the potato eelworm to hatch from its cysts.[11]

31   31a   32; 70% yield   33; solanoeclepin A

Four-Membered Rings by Ionic Reactions

The cyclobutene **14** used a few pages back in a cycloaddition was actually made by an ionic reaction from adipic acid **34**. Double bromination of the acid chloride and quenching with methanol gives **35** that cyclises[12] to **14** with NaH. Presumably one enolate is alkylated by the other bromide and then the second enolate eliminates **36** to give **14**.

34   35   14   36

Remarkable syntheses of polycyclic fused systems such as **38** by treatment of simple compounds **37** with base have emerged from Takasu and Ihara.[13] You might compare the skeleton of **38** with that of **30**.

37   38

Presumably the silyl enol ether of **37** adds in a conjugate fashion to the unsaturated ester **39** and the intermediate enolate then cyclises onto the cation **40** to give **38**. This will happen only if the stereochemistry of **40** is the same as that of the product **38** as the 4/5 and 4/6 ring fusions must both be *cis*. This suggests that the first step is reversible. The formation of the cyclobutane requires that particular relationship between ketone and unsaturated ester so this kind of reaction is less versatile than photochemical cyclisation. Asymmetric versions of these reactions are also known.[14] Probably the most versatile thermal method to make cyclobutanes uses ketenes and is the subject of the next chapter.

39; R = SiMe₃ → 40; R = SiMe₃ → 38

References

1. W. H. Perkin, *J. Chem. Soc.*, 1885, 801; 1886, 806; 1887, 1;   E. Haworth and W. H. Perkin, *J. Chem. Soc.*, 1894, 591;   *House*, pages 541–544.

2. Fleming, *Orbitals*, pages 86 and 208.

3. M. T. Crimmins, *Comp. Org. Synth.*, **5**, 123.

4. P. G. Bauslaugh, *Synthesis*, 1970, 287; M. T. Crimmins and T. L. Reinhold, *Org. React.*, 1993, **44**, 297.

5. D. C. Owsley and J. J. Bloomfield, *J. Chem. Soc. (C)*, 1971, 3445.

6. G. L. Lange, M.-A. Huggins and E. Neiderdt, *Tetrahedron Lett.*, 1976, 4409.

7. R. G. Salomon and S. Ghosh, *Org. Synth.*, 1984, **62**, 125.

8. S. W. Baldwin and J. M. Wilkinson, *Tetrahedron Lett.*, 1979, 2657.

9. Fleming, *Orbitals* page 219.

10. M. Fétizon, S, Lazare, C. Pascard and T. Prange, *J. Chem. Soc., Perkin Trans. 1*, 1979, 1407.

11. B. T. B. Hue, J. Dijkink, S. Kuiper, K. K. Larson, F. S. Guziec, K. Goubitz, J. Fraanje, H. Schenk, J. H. van Maarseveen and H. Hiemstra, *Org. Biomol. Chem.*, 2003, **1**, 4364.

12. R. N. MacDonald and R. R. Reitz, *J. Org. Chem.*, 1972, **37**, 2418.

13. K. Takasu, M. Ueno and M. Ihara, *J. Org. Chem.*, 2001, **66**, 4667.

14. K. Takasu, K. Misawa and M. Ihara, *Tetrahedron Lett.*, 2001, **42**, 8489.

33 Strategy XV: The Use of Ketenes in Synthesis

We met ketenes in chapter 31 where they were intermediates in the Arndt–Eistert chain extension procedure. Now we are going to take a wider view of their value in synthesis. A ketene **1** is very electrophilic at the curious sp carbon atom (marked * in **1**) and combines with nucleophiles **2** to give enolates that are protonated at carbon **3** to give acylated compounds **4**.

Ketenes are rarely isolated as they dimerise easily. Ketene itself **2** gives the lactone **5** but dimethylketene **6** gives the diketone **7**. Other ketenes may give either type of dimer. Only a few ketenes, such as diphenyl ketene, are normally isolated.

Ketenes are normally prepared by the base-catalysed elimination of HCl from an acid chloride **9** or by elimination of chlorine from a chloroalkyl acid chloride with zinc dust, often assisted by ultrasound. For reactions with nucleophiles, the solution would already contain the nucleophile before the ketene **6** was generated.

[2 + 2] Thermal Cycloadditions of Ketenes

Unlike ordinary alkenes, ketenes do 2 + 2 cycloadditions with themselves—the dimerisation above—and with other alkenes.[1] Reaction of dichloroketene with cyclobutadiene **11** to give the

Organic Synthesis: The Disconnection Approach. Second Edition   Stuart Warren and Paul Wyatt
© 2008 John Wiley & Sons, Ltd

dichloroketone **12** shows that they prefer 2 + 2 to 4 + 2 cycloadditions and also shows off the regioselectivity you would expect: the sp carbon reacts with the most nucleophilic end of the alkene.[2] The 'mechanism' **13** shows the major orbital interaction as the two reagents approach each other: the reaction may well be a concerted cycloaddition.

Reaction of dichloroketene with *cis* or *trans* cyclo-octene suggests that it is a concerted reaction; each gives stereospecifically a different stereoisomer of the adduct: *cis*-**15** gives *cis*-**16** while *trans*-**17** gives *trans*-**18**. The marked hydrogen atoms should make this clear. The very reactive *trans*-cyclo-octene **17** gives a 100% yield so there is no room for any **16** in the product.[3]

The disconnections for this reaction are, of course, of one of the two sets of opposite C–C bonds in the cyclobutanone. Each will give a ketene and an alkene and a choice has to be made. Adduct **12** is easy: disconnection **12a** gives two simple starting materials while disconnection **b**, though the cyclisation would be intramolecular, gives a starting material of some difficulty. For example, the middle alkene must be *cis* for the reaction to have a chance.

Others are not so obvious. Cyclobutanone **22** might come from diphenylketene and the diene **21** (disconnection **22a**) or from ketene itself and the diene **23** (disconnection **22b**). No doubt both dienes could be made but the regioselectivity of both cycloadditions looks in doubt.

The diene **23** would probably react at the less hindered alkene or at the wrong end of the more activated alkene with the two phenyl groups. But the chances of **21** reacting at the right end are much better: it is both the more nucleophilic and the less hindered alkene and the regioselectivity looks right too. This is in fact how it was made: diphenylketene **20** adds to the *cis*-diene **21** to give **22** in 99% yield.[4]

Rearrangements of the Products of Ketene 2 + 2 Cycloadditions

Both the Baeyer–Villiger and the Beckmann rearrangement are used on the cyclobutanones formed in these cycloadditions. Dechlorination of adduct **12** with zinc and rearrangement with a peroxy-acid gives a lactone **25** widely used in prostaglandin synthesis.[5] Note that the more substituted carbon migrates and does so with retention of configuration.

An example[6] with regioselectivity in both reactions is the cyclobutanone **27**. Addition of dichloroketene, that must be made by the zinc dehalogenation method for good results, to the cyclohexenone **26** gives just one isomer of **27** that can be dehalogenated (zinc again) and oxidised to the lactone **29**. Again the more substituted carbon atom migrates with retention.

The Beckmann rearrangement is used in a similar way to produce the lactam **32**, an intermediate in the synthesis of swainsonine **33**. Stereoselective addition of dichloroketene to the enol ether **30** gave one isomer (~95:5) of cyclobutanone **31**. Beckmann rearrangement with a sulfonated hydroxylamine and dechlorination gave the lactam **32** in 34% yield over five steps[7] from a precursor of **30**. Note that the *cis*-alkene **30** gives the *trans* cyclobutanone selectively.

Even the diazomethane ring-expansion (chapter 31) works well on cyclobutanones: they seem all too eager to lose their strain by ring expansion. This unusual sequence leads to the cyclopentenone **38** used for various conjugate additions.[8] Presumably CHMe migrates better than CCl_2 giving **36** and the zinc treatment removes just one chlorine atom. The LiBr/DMF treatment must eliminate via the enol(ate) of **37** giving the more substituted alkene[9] **38**.

Ketene Dimers as Acylating Agents

We mentioned the dimer of ketene **6** itself at the start of this chapter: it is a cyclic enol ether and a good acylating agent. Nucleophiles attack the carbonyl group **39** expelling the enolate **40** of the acetoacetyl derivative **41**. The disconnection is shown on **41** and the ketene dimer represents synthon **42**.

The heterocycle **43** was needed as an intermediate in a cytochalasan synthesis.[10] Disconnection of the 1,3-diCO relationship between the two ketones reveals the amide **44** that is the acetoacetyl derivative of phenylalanine ethyl ester **45**.

The synthesis is straightforward: the ester **45** combines with the ketene dimer **6** with catalysis by base providing the base is not in excess. Cyclisation to give a five-membered cyclic amide occurs with more base to give **43** that actually exists as the enol **43a**.

References

1. Fleming, *Orbitals*, page 143;   Clayden, *Organic Chemistry*, chapter 35.
2. R. W. Holder, *J. Chem. Ed.*, 1976, **53**, 81.
3. R. Montaigne and L. Ghosez, *Angew. Chem. Int. Ed.*, 1968, **7**, 221.
4. R. Huisgen and P. Otto, *Tetrahedron Lett.*, 1968, 4491.
5. E. J. Corey, Z. Arnold and J. Hutton, *Tetrahedron Lett.*, 1970, 307;   M. J. Dimsdale, R. F. Newton, D. K. Rainey, C. F. Webb, T. V. Lee and S. M. Roberts, *J. Chem. Soc., Chem. Commun.*, 1977, 716.
6. P. W. Jeffs, G. Molina, M. W. Cass and N. A. Cortese, *J. Org. Chem.*, 1982, **47**, 3871.
7. J. Ceccon, A. E. Greene and J.-F. Poisson, *Org. Lett.*, 2006, **8**, 4739.
8. A. E. Greene, J.-P. Lansard, J.-L. Luche and C. Petrier, *J. Org. Chem.*, 1984, **49**, 931.
9. A. E. Green and J. P. Deprés, *J. Am. Chem. Soc.*, 1979, **101**, 4003;   *J. Org.Chem.*, 1980, **45**, 2036.
10. T. Schmidlin and C. Tamm, *Helv. Chim. Acta*, 1980, **63**, 121.

34 Five-Membered Rings

Unlike three-, four- or six-membered rings, five-membered rings are often made by standard carbonyl chemistry. This is because five-membered rings are the easiest to make by carbonyl condensations as they have kinetic and thermodynamic advantages over open chain compounds (chapter 29). This chapter gives a selection of such methods and the next chapter looks at some special methods for making five-membered rings.

Five-Membered Rings from 1,4-diCarbonyl Compounds

Cyclopentenones **1** disconnect to 1,4-dicarbonyl compounds **2**. Any of the methods in chapter 25 may be used to make these but the regioselectivity of the cyclisation is an important consideration. If R = Me **3**, cyclisation can lead only to **1**; R = Me but, if R = Et **4**, could cyclise to **5** or **6** depending on which ketone forms the enolate. Thermodynamically **6** is favoured as it has a more highly substituted alkene, but it is close.

A simple example is the cyclopentenone **7** because the keto-aldehyde **8** can cyclise only one way as the aldehyde cannot enolise. The best 1,4-dicarbonyl disconnection is probably **8** giving some enolate equivalent **10** of isobutyraldehyde and a reagent for the unnatural synthon **9** such as the bromoketone **11**.

Organic Synthesis: The Disconnection Approach. Second Edition   Stuart Warren and Paul Wyatt
© 2008 John Wiley & Sons, Ltd

In fact the workers who wanted **7** for photochemical addition to alkenes (chapter 32) chose[1] to use propargyl bromide **14** and an enamine **13** of the aldehyde **12**. Mercury-catalysed hydration of **15** gave **8** which cyclised to **7** in base.

Some molecules are studied for their theoretical interest: one being cyclopentadienone **16**. But it turns out that this dimerises instantly by a Diels-Alder reaction and cannot be studied. The simplest cyclopentadienone that can be made is the tetraphenyl compound **17**. Aldol disconnection gives **18** but we can now do a second aldol disconnection to reveal the two symmetrical starting materials **19** and **20**.

Benzil **20** can be made by the oxidation of benzoin[2] **21** (chapter 23) and it combines with **19** in one step under base catalysis[3] without the need to isolate **18**. The problem with these compounds is that **16** has only four π-electrons delocalised round the ring and is anti-aromatic. Clearly four phenyl groups help stability but **17** exists as deep purple crystals showing an unusually small gap between the populated and unpopulated orbitals.

Cyclopentyl Ketones from 1,6-diCarbonyl Compounds

We have already seen (chapter 19) the synthesis of cyclopentanone itself **24** via the useful β-ketoester **23** from adipate esters **22**. In the same way unsaturated ketones **25** disconnect with an aldol in mind to the 1,6-dicarbonyl compound **26**. There may again be regioselectivity questions in the cyclisation.

Thus both unsaturated carbonyl compounds **27** and **30** disconnect to the same 1,6-dicarbonyl compound **28** that reconnects to natural limonene **29**. There are two chemo-selectivity problems here: how do we cleave one alkene in limonene without cleaving the other and how do we control the cyclisation?

It turns out that epoxidation prefers the more substituted alkene in the ring. The epoxide **31** can then be opened to the diol **32** and cleaved with periodate to give **28**. The ketoaldehyde **28** was not isolated but cyclised immediately.[4]

But what about this cyclisation? In stronger protic bases like KOH in water, all the cyclisations are reversible and the more stable ketone **27** is formed by thermodynamic control. In buffered conditions (weak amine base and weak acid) only the more reactive aldehyde enolises and **30** is formed by kinetic control.

Cyclopentanes from 1,5-diCarbonyl Compounds

The silicon modification of the acyloin condensation gives excellent yields of five-membered rings. The simple spiro compound **35** provides a perfect illustration. The traditional acyloin without silicon gives a paltry 18% yield: with Me_3SiCl present, the yield (of **33**) rises[5] to 87%.

Even with a spiro four-membered ring, the reaction works well. In a synthesis of the theoretically interesting molecule **39**, the silicon acyloin and its hydrolysis gave good yields.[6]

However, the presence of an alkene *exo* to the chain **40** stops the reaction, presumably because the 120° angle holds the ends too far apart. The solution is conjugate addition of an amine **41**: the acyloin then works well **42** and the synthesis of the flavouring compound 'corylone' **43** is completed simply by a silica column.[7] Hydrolysis of the silyl enol ethers leads to elimination of Me_2NH under the slightly acidic conditions.

Synthesis of Cyclopentanes by Double Sequential Conjugate Addition

In chapter 21 we saw how a conjugate addition, followed by an aldol condensation, gave six-membered rings. Now we can see that conjugate addition followed by another conjugate addition can give five-membered rings. The starting material **45** is easily made by alkylation of malonate with the allylic halide **44**. Treatment with base and an unsaturated ketone **46** gives a cyclopentane with high (>50:1) stereoselectivity in favour of the *trans* compound[8] **47**.

The conversion of **45** and **46** into **47** occurs in one operation. Evidently the anion of **45** adds **48** to the enone **46** and the enolate produced cyclises by a second conjugate addition **49**. This type of reaction gives a more highly substituted cyclopentane than we have seen so far.

The disconnection is of the two 1,5-diCO relationships present in **47**: it doesn't matter (much) which you do first: the second follows. Disconnection **47a** leads us straight back to **51** and hence our starting materials but **47b** needs a little more imagination to see the second disconnection on **50**. This sequence leads to a five-membered ring because there is only one CH_2 group between the bromine and the alkene in **44**. If there were two, a six-membered ring would be formed. That is the subject of chapter 36.

References

1. S. Wolff, W. L. Schreiber, A. B. Smith and W. C. Agosta, *J. Am. Chem. Soc.*, 1972, **94**, 7797; P. D. Magnus and M. S. Nobbs, *Synth. Commun.*, 1980, **10**, 273.

2. *Vogel*, page 1045.

3. *Vogel*, page 1101.

4. J. Meinwald and T. H. Jones, *J. Am. Chem. Soc.*, 1978, **100**, 1883; J. Wolinsky and W. Barker, *J. Am. Chem. Soc.*, 1960, **82**, 636; J. Wolinsky, M. R. Slabaugh and T. Gibson, *J. Org. Chem.*, 1964, **29**, 3740.

5. J. J. Bloomfield, D. C. Owsley and J. M. Nelke, *Org. React.*, 1976, **23**, 259.

6. R. D. Miller, M. Schneider and D. L. Dolce, *J. Am. Chem. Soc.*, 1973, **95**, 8468.

7. R. C. Cookson and S. A. Smith, *J. Chem. Soc., Perkin Trans. 1*, 1979, 2447.

8. R. A. Bunce, E. J. Wamsley, J. D. Pierce, A. J. Shellhammer and R. E. Drumright, *J. Org. Chem.*, 1987, **52**, 464.

Strategy XVI: Pericyclic Reactions in Synthesis: Special Methods for Five-Membered Rings

35

Background Needed for this Chapter Reference to Clayden, *Organic Chemistry:* Chapter 36: Pericyclic Reactions II: Sigmatropic and Electrocyclic Reactions.

The only pericyclic reactions we have used so far have been cycloadditions: the Diels-Alder reaction in chapter 17 and 2 + 2 cycloadditions in chapter 33. Electrocyclic and sigmatropic reactions are also used in synthesis and, as each is the basis for a synthesis of five-membered rings, they are grouped together here.

Electrocyclic Reactions

An electrocyclic reaction is the formation of a new σ-bond across the ends of a conjugated π-system or the reverse. They thus lead to the creation or destruction of one σ-bond. Hexatrienes **1** can cyclise to six-membered rings **2** in a disrotatory fashion but we shall be more interested in versions of the conrotatory cyclisation of pentadienyl cations **3** to give cyclopentenyl cations **4**. The different stereochemistry results from the different number of π-electrons involved.[1]

The Nazarov Reaction

The Nazarov[2] is probably the most important of reactions like **3**. The cation **6** is formed from a dienone **5** by protonation and cyclises to the allylic cation **7**. Though this is presumably a conrotatory process, the stereochemistry is usually lost in the formation of the cyclopentenone **9**.

Organic Synthesis: The Disconnection Approach. Second Edition   Stuart Warren and Paul Wyatt
© 2008 John Wiley & Sons, Ltd

Thus the natural product damascenone **10**, responsible in part for the smell of roses, cyclises in acid to the cation **11** that can lose a proton from one side only to give[3] **12**. The disconnection for the Nazarov reaction is of the single bond in the five-membered ring opposite the carbonyl group **12a**.

The double bond in the ring can be part of a benzene ring so that Nazarov disconnection of **13** reveals an aromatic ketone that can surely be made by a Friedel-Crafts reaction on some derivative of the acid **16**.

You might think of using the acid chloride **16**; X = Cl with a Lewis acid and that might well be successful, but it turns out that reaction of the acid **16**; X = OH with the ether **15** using polyphosphoric acid as catalyst does both the Friedel-Crafts and the Nazarov in one step. The yield is only 70% but it is a very short synthesis.[4]

Aromatic heterocyclic compounds such as the *N*-tosyl pyrrole **17** can also be used with anhydride catalysis (acid might destroy the pyrrole). Regioselectivity is determined by acylation next to the nitrogen[5] and cyclisation follows.[6]

In other situations, Lewis acids such as AlCl₃ may be better. Disconnection of the tricyclic compound **21** is best at the middle ring and Nazarov is ideal as it gives a simple dienone **22**. Though **22** has two enones, both are in rings so we should rather disconnect a bond between rings to give the synthons **23** and **24**.

One reason for choosing this disconnection is because the lithium derivative **26** of dihydropyran **25** is easily made and the enal **27** is the product of cyclisation of hexadial. Oxidation of **28** (MnO$_2$ is good for the oxidation of allylic alcohols) and treatment with AlCl$_3$ gives **21** in high yield. The remaining alkene in **21** is conjugated with the ether oxygen and the stereochemistry is inevitably *cis* for two fused five-membered rings.[7]

Sigmatropic Rearrangements

A sigmatropic rearrangement is a unimolecular reaction **29** in which a σ-bond moves from one place in a molecule to another. In these reactions there is no net change in the number of σ-bonds. They are classified by numbering both ends of the old σ-bond '1' **29a** and numbering round in both directions to find the new σ-bond **30a**. So this reaction **29** is a [3,3]-sigmatropic rearrangement. The sum of the two numbers gives the size of the cyclic transition state[8] **31**.

This reaction does not create a ring but there is an important group of sigmatropic rearrangements that make five-membered rings—the vinyl cyclopropane to cyclopentene rearrangement.

The Vinyl Cyclopropane to Cyclopentene Rearrangement

On heating, vinyl cyclopropanes **32** isomerise to cyclopentenes[9] **33**. This is a [1,3]-sigmatropic rearrangement of **32a** to **33a** having a rather strained four-membered cyclic transition state **34**. It usually requires strong heating, typically >300 °C. There is disagreement about the mechanism: some people think it is concerted **32** and others that a C–C bond in the three-membered ring splits **32b** to give a diradical **35** that can reunite to give **33**.

So the cyclopropane **36** isomerises at high temperature, no doubt via the [1,3] shift to **37**, formation of the extended enol **38** and movement of the alkene into conjugation follows. The product **39** was used in a synthesis of the natural product zizaene.[10]

The disconnection looks tricky but it is all right if you simply reverse the rearrangement, by drawing the mechanism for the imaginary reverse reaction. There may well be two possible starting materials. Thus cyclopentene **40**, needed for a photochemical experiment, could be disconnected as **40a** or **40b**. There is no obvious way to continue from **41** but **42** has an enone that could be made from aldehyde **43** and some reagent for the enolate of acetone **44**.

The aldehyde was made by alkylation of the nitrile **45** and reduction (actually using LiAlH$_4$, though we might prefer DIBAL), an HWE olefination and conversion of the ester **46** into the ketone **42**. Finally, the rearrangement[11] required 400 °C.

If there are heteroatoms in the molecule, the [1,3] shift can be catalysed by acids or Lewis acids. The heterocycle **49**, needed for a synthesis of alkaloids from narcissi, is formed from the cyclopropyl imine **48** with HBr catalysis.[12] The imine **48** comes from an aldehyde **47** made in the same way as **43**.

Lower temperatures are enough with a strong Lewis acid like Et$_2$AlCl. The cyclopropane **52** comes from available dihydrofuran **50** by rhodium-catalysed carbene insertion. Rearrangement at very low temperatures gives the cyclopentene **53** that actually has three five-membered rings fused together.[13]

[3,3]-Sigmatropic Rearrangements

We used the all-carbon Cope rearrangement **29** to introduce this section but now we want to feature the more useful Claisen rearrangements.[14] The aliphatic Claisen **54** works for most substituents because an alkene is lost and a much more stable carbonyl group is formed **55**. It doesn't matter whether we have an aldehyde (X = H), a ketone, (X = R), an acid (X = OH), an ester (X = OR) or an amide (X = NR$_2$), the reaction works well. The original Claisen rearrangement was the aromatic version **56** that gives an unstable non-aromatic intermediate **57** that quickly loses a proton to restore the aromatic ring and the product is a phenol **58**.

The disconnection for the aromatic Claisen is to reverse the rearrangement. This is a little simpler than those we have seen so far as one C–C bond is broken **59** and one C–O bond made. But you must remember to turn the allylic system back to front. This is easily seen if the starting material is drawn as **59a** with the dotted line representing a reconnection. The rest is a normal ether disconnection.

The allylic halide needed for the alkylation is easily made by some aldol or Wittig-style reaction to give **63** followed by reduction and conversion of OH to Br. Alkylation of phenols (pK_a 10) with allylic halides is very easy as a weak base such as carbonate is enough and the Claisen rearrangement merely requires heating.[15]

The aliphatic Claisen rearrangement is simpler in that there is no rearomatisation at the end but there is an ionic step first as the vinyl ether **67** has to be made and the easiest way to do that is from the allylic alcohol **65** by acetal exchange with another vinyl ether to give **66** and elimination to give **67**. All these steps, including the rearrangement occur under the same conditions[16] and the product is a γ,δ-unsaturated carbonyl compound **68**.

The easiest way to see that a Claisen rearrangement might be useful is to look at the relationship between the alkene and the carbonyl group. A simple C–C disconnection **69** suggests that allylation of an enolate **71** with allylic halide **70** would be a good method. But **70** will react at the wrong end of the allylic system so we need a method that turns the allylic system inside out and that is the Claisen rearrangement using **72** as the allylic alcohol.

Just as **65** gave **68**, so **72** gives **69** except that we must work at a higher oxidation level, using an orthoester MeC(OEt)$_3$ to make the ketene acetal **74** and hence the ester[17] **69** rather than the aldehyde **68**.

Stereoselectivity of the Claisen rearrangement

If the new alkene has geometry, the Claisen rearrangement is very *E*-selective. Thus the allylic alcohol **75** gives just the *E*-unsaturated aldehyde **77**. The transition state for the rearrangement is a six-membered ring **78** and has a chair conformation **79**. All the substituents are hydrogen except for R which will prefer an equatorial position. If you look at the right-hand side of this diagram **79** you will see that the framework of the *E*-alkene is already in that conformation.

A good illustration is the synthesis[18] of *E*-**84** by Claisen rearrangement of **82** (R is a protecting group). Notice that neither a carbonyl group nor a furan ring interferes with the reaction. The next step in this synthesis of porphobilinogen, the porphyrin precursor, was an intramolecular Diels-Alder reaction between the new alkene and the furan ring.

An important reason for making new molecules is to use them in the construction of organic materials such as highly branched polymers. The tetrabromo compound **89** is one of these and remarkably it can be made by a Claisen rearrangement.[19] The allylic alcohol **85** is made by the usual Wittig-reduction sequence and Claisen rearrangement with ethyl orthoacetate gives the unsaturated ester **86**. Hydroboration now gives the lactone **67**, reduction the diol **88**, and all the oxygens are then transformed into bromides in a single operation. This compound was used to make tetra-amines and hence 'dendrimers'—highly branched tree-like polymers.

The Claisen rearrangement can also be used to make amides if the dimethylacetal of DMF **91** is used to make the vinyl ether. So our final example is an introduction to the next chapter on six-membered rings where we shall use Birch reduction as a method of reducing aromatic compounds. Allylic alcohol **90** is made by Birch reduction of the aromatic compound and the Claisen rearrangement **92** has an extra feature. The rearrangement occurs across the top face of the ring so that the product **93** is a single diastereomer. You will notice that the other alkene moves into conjugation with the ester. This product was used in the synthesis of an alkaloid.[20]

References

1. Fleming, *Orbitals* page 103.
2. K. L. Habermas, S. E. Denmark and T. K. Jones, *Org. React.*, 1994, **45** 1; S. E. Denmark, *Comp. Org. Synth*, **5**, 751.
3. G. Ohloff, K. H. Schulte-Elte and E. Demole, *Helv. Chim. Acta*, 1971, **54**, 2913.
4. T. R. Kasturi and S. Parvathi, *J. Chem. Soc., Perkin Trans. 1*, 1980, 448.
5. Clayden, *Organic Chemistry*, chapter 43.
6. C. Song, D. W. Knight and M. A. Whatton, *Org. Lett.*, 2006, **8**, 163.
7. G. Liang, S. N. Gradl and D. Trauner, *Org. Lett.*, 2003, **5**, 4931.
8. Clayden, *Organic Chemistry*, chapter 36; Fleming, *Orbitals*, page 98; R. K. Hill, *Comp. Org. Synth.*, **5**, 785.
9. T. Hudlicky, T. M. Kutchan and S. M. Naqvi, *Org. React.*, 1985, **33**, 247.
10. E. Piers and J. Banville, *J. Chem. Soc., Chem. Commun.*, 1979, 1138.
11. H.-U. Gonzenbach, I.-M. Tegmo-Larsson, J.-P. Grossclaude and K. Schaffner, *Helv. Chim. Acta*, 1977, **60**, 1091; H. Künzel, H. Wolf and K. Schaffner, *Helv. Chim. Acta*, 1971, **54**, 868.
12. C. P. Forbes, G. L. Wenteler and A. Weichers, *Tetrahedron*, 1978, **34**, 487.

13. H. M. L. Davies, N. Kong and M. R. Churchill, *J. Org. Chem.*, 1998, **63**, 6586.

14. A. Martín Castro, *Chem. Rev.*, 2004, **104**, 2939; P. Wipf, *Comp. Org. Synth.*, **5**, 827.

15. D. S. Tarbell, *Org. React.*, 1944, **2**, 1.

16. R. Marbet and G. Saucy, *Helv. Chim. Acta*, 1967, **50**, 2095; A. W. Burgstahler and I. C. Nordin, *J. Am. Chem. Soc.*, 1961, **83**, 198.

17. Y. Nakada, R. Endo, S. Muramatsu, J. Ide and Y. Yura, *Bull. Chem. Soc. Jpn.*, 1979, **52**, 1511.

18. P. A. Jacobi and Y. Li, *J. Am. Chem. Soc.*, 2001, **123**, 9307.

19. K. S. Feldman and K. M. Masters, *J. Org. Chem.*, 1999, **64**, 8945.

20. T.-P. Loh and Q.-Y. Hu, *Org. Lett.*, 2001, **3** 279.

36 Six-Membered Rings

There are three general methods of making carbocyclic six-membered rings and each produces rings with a characteristic substitution pattern. The first uses carbonyl condensations and the best of these is the *Robinson annelation*[1] (chapter 21). The disconnections are aldol **1** and conjugate (Michael) addition **2**. The target molecule is a conjugated cyclohexenone.

The second method is the *Diels-Alder reaction* (chapter 17). The target molecule **5** also has a carbonyl group and an alkene but now only the alkene is in the ring. The carbonyl group is outside the ring and remote from the alkene. The simplest way to do the disconnection is to draw the mechanism of the imaginary reverse reaction **5a**. Start your arrows on the alkene and go whichever way round the ring you prefer **5a** or **5b**.

The third is partial or total *reduction of an aromatic ring*. Any catalogue lists a vast number of available substituted benzene rings. Saturated compound **8** can obviously be made by total reduction of **9** but it may not be obvious that partial reduction (Birch) allows the enone **11** also to be made from **9**. Birch reduction is the only new method here so we shall revise the Robinson and the Diels-Alder and concentrate on Birch.

Organic Synthesis: The Disconnection Approach. Second Edition   Stuart Warren and Paul Wyatt
© 2008 John Wiley & Sons, Ltd

Carbonyl Condensations: The Robinson Annelation

Tremendous improvements have been made in the Robinson annelation in recent times: organic catalysts have been developed, some giving single enantiomers of products (see *Strategy and Control*) and some, such as **12**, giving very fast reactions and complete control[2] over the stereochemistry of the aldol intermediate **13**.

Neither starting material need be cyclic. Combination of the acyclic enone **14** and β-ketoester **15** with an amine catalyst gives high yields of **16** and good (>97:3) stereoselectivity.[3] Both **13** and **16** can easily be dehydrated to the enones **1** or **17**.

Other Ionic Cyclisations

The Robinson annelation is by no means the only ionic reaction that makes six-membered rings. Six-membered rings form easily so trapping a Nazarov intermediate (chapter 35) makes good sense. The Friedel-Crafts-like disconnection **18** suggests a most unlikely cation **19** until we realise that it would be formed in the Nazarov cyclisation of the dienone **20** whose synthesis is discussed in the workbook.

The catalyst used was TiCl₄ and the complex **21** cyclises in a conrotatory fashion so that the two Hs end up *trans* in the intermediate **22**. Cyclisation to the activated *para* position of the benzene ring that is already attached to the bottom face of the five-membered ring gives the titanium enolate **23** and protonation puts the ethyl group in the more favourable 'down' position, *anti* to the nearer methyl group.[4]

A similar cyclisation of an alkene derived from geranyl acetate **24** by dihydroxylation and formation of the epoxide **26** leads to a substituted cyclohexane **28**. The Lewis acid ZrCl$_4$ is used to open the epoxide and the alkene attacks intramolecularly **27** to give eventually the *syn*-compound **28** with both substituents equatorial. The alignment of the alkene and the epoxide in a chair conformation **27a** is responsible for the diastereoselectivity.[5] Note the regioselectivity: the less substituted end of the alkene attacks the more substituted end of the epoxide **27**. These are just two examples of the very many ordinary ionic reactions that can be used to make six-membered rings.

The Diels-Alder Reaction

We established in chapter 17 that the Diels-Alder reaction offers exceptional control over all forms of selectivity and so it is one of the most important of all reactions used in synthesis.[6] So when Nicolaou needed **29** for a synthesis of columbiasin A, he was prepared to go to some lengths to change **29** into a Diels-Alder substrate. It looks unpromising: neither ring has the alkene nor is there a carbonyl group in the right position. But the ketone in ring A could come from an enol ether **30** and the benzene ring **B** could come from a quinone by the reverse logic.

Making the Diels-Alder disconnection by drawing the mechanism of the reverse reaction **31a** gives a new enol ether **32** and a quinone **33**. The enol ether **32** is a derivative of the simple enone **34** and can be made by trapping the kinetic enolate with a suitable silyl group.

Nicolaou preferred[7] to use the *t*-BuMe$_2$Si group for 'R' in **32** and found that the Diels-Alder with **32a** went with complete regio- and stereo-control. The stereocontrol is not important as the next step will destroy it but the regiocontrol is important and interesting. The OMe group on the quinone is conjugated with the lower carbonyl group so the *'para'* orientation is between the *t*-BuMe$_2$SiO group and the other carbonyl group. For the same reason the right-hand alkene in the quinone is less electrophilic. It is remarkable that these small factors have such a big effect.

Aromatisation to **29** simply requires enolate formation and methylation to give **30**. Hydrolysis of the silyl enol ether with CF$_3$CO$_2$H gave the ketone **29**.

A Synthesis of the Guanacastetepene Skeleton

The guanacastepenes are antibiotics from a rare Costa Rican fungus and have the basic skeleton **35**. This has five-, six- and seven-membered carbocyclic rings but there is a hint of Diels-Alder possibilities about it in that ring **C** is a cyclohexene. In fact Shipe and Sorensen[8] chose **36** as a key intermediate but no direct Diels-Alder disconnection looks likely. Does it help if we reconnect to the lactone **37**?

It does because **37** can be made by a Baeyer–Villiger rearrangement from the ketone **38** as the more substituted centre migrates with retention. We have the carbonyls outside the ring (the two CO$_2$Me groups) but we cannot have the carbonyl group on the ring. So we change it to an enol ether **39** and Diels-Alder disconnection reveals two simple starting materials **40** and **41**.

The diene **41** was made from the cyclohexenone **42** (easily made by intramolecular aldol) by kinetic enolate formation and silylation. Diels-Alder reaction followed by hydrolysis of the intermediate **39**; R = SiMe$_3$ gave the ketone **38** and Baeyer–Villiger oxidation gave the lactone **37**. Notice the very high yields, especially in the rearrangement, leaving no doubt as to the total control over selectivity. Opening of the lactone **37** in acidic methanol gave the key intermediate **36** with ring **C** complete. The original six-membered ring in **42** has disappeared to be replaced by a new six-membered ring formed by the Diels-Alder reaction.

Reduction of Aromatic Compounds

Total reduction of a benzene ring requires pressure and active catalysts and is more easily done industrially than in the lab. We might choose such a method if substituents that are unhelpfully related in an aliphatic compound such as **43** and **45** may be conveniently related in the aromatic equivalents **44** and **46**. This usually means a '*para*' relationship in the target molecule.

The phenol **44** can obviously be made by a Friedel-Crafts reaction and the amine **45** by reduction of the nitro group in **46** as well as of the benzene ring. Since OH and OEt are both *o*, *p*-directing, the syntheses are simple.[9]

A more interesting case is the antispasmodic drug dicyclomine **50** with its two six-membered rings. Removing the ester group **51** makes it easier to see that one ring could be derived from a benzene ring. If we also change the acid to a nitrile **52**, the other six-membered ring can be made by alkylation. Ring **B** cannot be made by reduction as it has a quaternary centre.

The synthesis is straightforward. The nitrile **53** is alkylated and treated directly with acidic
ethanol to give the ester **54** so that the new ester **55** can be made by ester exchange. The reduction
of the benzene ring is the last step.[10]

Partial (Birch) Reduction of Benzene Rings

Birch reduction[11] is the partial reduction of aromatic rings by solvated electrons produced when
alkali metals dissolve (and react) in liquid amines. Typical conditions are sodium in liquid
ammonia or lithium in methylamine. These electrons add to benzene rings to produce, probably, a dianion **57** that is immediately protonated by a weak acid (usually a tertiary alcohol)
present in solution. The anions in the supposed intermediate **57** keep as far from each other as
they can so the final product is the non-conjugated diene **58**. It is important to use the blue
solution of solvated electrons before it reacts to give hydrogen and NaNH$_2$.

If a benzene ring substituted with an electon-donating group (typically R or RO) is reduced in
this way, the dianion now keeps away from the electron-donating group **60** so that the product
from anisole **59** is the enol ether **61**. Hydrolysis under mild conditions gives the unconjugated
enone **62**, though more vigorous conditions move the alkene into conjugation.

On the other hand, an electron-withdrawing substituent, particularly a carbonyl group **63**, will
attract the anion **64** and again we get a non-conjugated product. If the acid **63**; R=H is used, the
less stable, non-conjugated anion in the intermediate **64**; R=H captures the proton from the acid
giving the enolate dianion **66** as the immediate product. This can be combined with electrophiles
such as alkyl halides as we shall see.

The epoxide **67** obviously comes from the diene **68** and hence by Birch reduction from **69**. In
the reduction **68** has both anions away from the alkyl substituents but the alternative **70** would
be less stable as the dianion.

Sodium in liquid ammonia does the reduction and the more substituted, and hence more nucleophilic, alkene reacts with the peroxyacid to give the target molecule.[12] Peroxyphthalic acid **71** was used as the oxidant.

An Alkaloid Synthesis

Guillou and her team[13] needed **73** for a synthesis of the cytotoxic alkaloid maritidine **72**. Reductive amination looks a good bet with the amine **74** combining with the aromatic aldehyde.

The amine **74** certainly looks like a product of Birch reduction of the simple *para*-disubstituted benzene **76** via the enol ether **75**.

It turned out that the Birch reduction of **76**; R=Me had already been done in 1958 by the standard sodium in liquid ammonia procedure but gave a low yield (~20%) of the conjugated ketone as the first compound isolated.[14] Guillou and her team improved on this by using lithium and *t*-BuOH at low temperature to give **78** in nearly quantitative yield.[15] The reductive amination needed only NaBH₄ in MeOH at room temperature and gave 78% of the enol ether of **73** from which the synthesis of **72** was completed.

Reduction of Aromatic Carboxylic Acids

In the last chapter we used **79** in an aliphatic Claisen rearrangement to make an anisatin inter-mediate. You can now see that **79** was made by Birch reduction of **80**. With a Friedel-Crafts in mind, we change the alcohol for a ketone **81** and start with **82**.

This plan looks good but in fact the CO_2Me group proved incompatible with the reactions so they made the alcohol **86** without it. Using PPA for the Friedel-Crafts reaction meant that the methyl ester **84** could be cyclised directly. The large reducing agent preferred to approach the flat five-membered ring from the side opposite the methyl group.

The next step is a directed lithiation—this subject is treated in *Strategy and Control*—to give the acid **88** via the lithiated species **87**. Birch reduction moves the two alkenes next to the electron-donating group (the three alkyl groups) and away from the electron-withdrawing group[16] (CO_2H) **79**. The rest of the synthesis appears in chapter 35.

Birch reduction of acid derivatives is even more productive if the first-formed enolate (as **66**) is used as a nucleophile. Our example also links this chapter with the last as a Cope rearrangement is featured. A group of alkaloids including mesembrine have the bicyclic structure **89**. Removing the structural nitrogen atom by standard disconnections (chapters 6 and 8) leaves the carbon skeleton **91** that does not immediately look a Birch reduction product.

The aromatic precursor is **92** reduced and then alkylation of the enolate anion **93** gives **94**, which was hydrolysed in aqueous HCl to the ketone **95**. In fact the 'NR$_2$' group was chiral and **95** was indeed one enantiomer.[17] So we have added a three-carbon side chain but in the wrong place.

The allyl group is moved to the right position **96** on heating in 1,2-dichlorobenzene by a Cope rearrangement. This occurs because the conjugation of the alkene with the two carbonyl groups in **96** is more effective than that with the aryl group in **95**. Ozonolysis liberates the aldehyde **97** and the reductive amination is followed by spontaneous Michael addition to close the five-membered ring **98**. All that remains is to remove the amide group from the β-ketoamide **98** by hydrolysis and decarboxylation.

Each of these methods of making six-membered rings has its own special characteristics and taken together they are versatile and powerful. They are not the only ways to make six-membered rings but they are the ones you should consider first for any new problem.

References

1. M. E. Jung, *Tetrahedron*, 1976, **32**, 3;   R. E. Gawley, *Synthesis*, 1976, 777.

2. S. G. Davies, R. L. Sheppard, A. D. Smith and J. E. Thompson, *Chem. Commun.*, 2005, 3802.

3. N. Halland, P. S. Aburel and K. A. Jørgensen, *Angew. Chem. Int. Ed.*, 2004, **43**, 1272.

4. C. C. Browder, F. P. Marmsätter and F. G. West, *Org. Lett.*, 2001, **3**, 3033.

5. M. Bovolenta, F. Castronovo, A. Vadalà, G. Zanoni and G. Vidari, *J. Org. Chem.*, 2004, **69**, 8959.

6. W. Oppolzer, *Comp. Org. Synth.*, **5**, 315;   W. R. Roush, *Comp. Org. Synth.*, **5**, 513; E. Ciganek, *Org. React.*, 1984, **32**, 1.

7. K. C. Nicolaou, G. Vassilikogiannakis, W. Mägerlein and R. Kranich, *Angew. Chem., Int. Ed.*, 2001, **40**, 2482.

8. W. D. Shipe and E. J. Sorensen, *Org. Lett.*, 2002, **4**, 2063.

9. S. Winstein and N. J. Holness, *J. Am. Chem. Soc.*, 1955, **77**, 5562;   E. L. Eliel, R. J. L. Martin and D. Nasipuri, *Org. Synth. Coll.*, 1973, **5**, 175;   R. W. West, *J. Chem. Soc.*, 1925, 494.

10. C. H. Tilford, M. G. van Campen and R. S. Shelton, *J. Am. Chem. Soc.*, 1947, **69**, 2902.

11. *Vogel*, page 1114;   P. W. Rabideau and Z. Marcinow, *Org. React.*, 1992, **42**, 1;   L. N. Mander, *Comp. Org. Synth.*, **8**, 489.

12. E. Giovanni and H. Wegmüller, *Helv. Chim. Acta*, 1958, **41**, 933.

13. C. Bru, C. Thal and C. Guillou, *Org. Lett.*, 2003, **5**, 1845.

14. C. B. Clarke and A. R. Pinder, *J. Chem. Soc.*, 1958, 1967.

15. C. Guillou, N. Milloy, V. Reboul and C. Thal, *Tetrahedron Lett.*, 1996, **37**, 4515.

16. T.-P. Loh and Q.-Y. Hu, *Org. Lett.*, 2001, **3**, 279.

17. T. Paul, W. P. Malachowski and J. Lee, *Org. Lett.*, 2006, **8**, 4007.

37 General Strategy C: Strategy of Ring Synthesis

This chapter collects ideas from the last eight chapters on ring synthesis and puts them into the context of our general approach to strategy. No grand new principles are needed as we shall use the same guidelines already established in chapters 11 and 28 with one or two extra guidelines for cyclic compounds.

Cyclisation to Control Selectivity

Cyclisations are easy. In chapters 7 and 21 we saw that the type of control needed to make open chain compounds is often unnecessary for cyclisations as intramolecular reactions usually take precedence over intermolecular. If, therefore, a difficult step needs to be used in a synthesis, it is good strategy to make it a cyclisation.

Corey[1] wanted to make the ketone **2** as an intermediate in the synthesis of the marine allomone **1** (a substance exuded by an organism and used by a predator). Disconnection **2a** (Friedel-Crafts alkylation) would be easy to realise as it occurs *para* to the powerful *o,p*-directing OMe group. But disconnection **2b** is more difficult as it must occur *meta* to OMe. So we should make the formation of bond **b** a cyclisation and disconnect it *first*.

The position of the branchpoint in **3** suggests a C–C disconnection with conjugate addition in mind and we can choose between **3a** and **3b**.

The synthesis starts with *ortho*-cresol **6** which is methylated and the easy substitution *para* to the OMe group puts in the aldehyde by a Vilsmeier acylation. A Knoevenagel-style aldol

---

Organic Synthesis: The Disconnection Approach. Second Edition   Stuart Warren and Paul Wyatt
© 2008 John Wiley & Sons, Ltd

gives the unsaturated acid **4** and copper-catalysed addition of the Grignard to the methyl ester **9** completes the carbon skeleton. Cyclisation with PPA occurs at the less substituted position *para* to the methyl group.

Early Disconnection of Small Rings

It is generally good strategy to disconnect a small (three- or four-membered) ring at an early stage or at least consider how the small ring might be made. The special methods needed for small rings often dominate the strategy. So for compound **11**, having three- and six-membered rings and two ketones, one protected as an acetal, carbene addition to the alkene **12** looks a good bet and this should immediately suggest a Birch reduction for the six-membered ring.

The methyl ether was chosen for **14** and it turned out that the enol ether **13** could be converted directly to the acetal **12**. It is essential to have one ketone protected before the other is introduced. The diazoketone (with copper catalysis) was used to put in the three-membered ring.[2]

Four-Membered Rings and New Reagents for Given Synthons

The tricyclic ketone **15** with four-, five- and six-membered rings must be made by a 2 + 2 photochemical cycloaddition (chapter 32), the small ring again dominating the strategy. There are two such disconnections **15a** and **15b**, each starting material **16** and **17** having an enone and an isolated alkene. We need not be concerned about the regio- or the stereoselectivity as only one cyclisation is possible in each case.

There is no immediately obvious disconnection of **17** but there is a strategic bond in **16** whose disconnection would make for a short synthesis. Now we have a new problem: neither polarity of disconnection **16a** is entirely satisfactory. We can easily think of reagents for **18** (Grignard reagent) or **20** (alkyl halide) but what about **19** and **21**? Yet we persist with this strategy because the bond between the ring and the chain is *strategic*.

At least **19** has the natural polarity of the enone but conjugate addition would lose the alkene so we need to add a leaving group at the site of the plus charge **22**. One possibility is an enol ether **23** as these are easily made from 1,3-diketones **24** and a suitable alcohol.

Ethanol was used to make the enol ether **23**; R = Et, the addition of the Grignard **25** was successful and the cycloaddition went in excellent yield.[3]

A more challenging example is the isomeric ketone **26**. The same two disconnections reveal two enones **27** and **28**. This time the regioselectivity of the 2 + 2 cycloadditions is 'wrong' in both cases but again we are not unduly concerned as they will be intramolecular.

We have to choose between making a five-membered ring **28** or a six-membered ring **27** and we shall choose the latter since it offers more possibilities. Disconnection at the branchpoint gives synthon **20** that can simply be an alkyl halide. The nucleophilic synthon **29** could be an enolate providing we move the alkene out of conjugation **30** and this leads back to a Birch reduction product.

Experiments showed[4] that the free acid **32** could be used for the Birch reduction and the product alkylated without work-up. Treatment with aqueous HCl hydrolysed the vinyl ether, decarboxylated the resulting β-ketoacid, and moved the alkene into conjugation **27**. As expected, the photochemical cycloaddition gave a high yield of the right positional isomer[5] **26**. This chemistry was used by Mander in his synthesis of gibberellic acid.[6]

Alternative Disconnections

We should emphasise that general guidelines on strategy suggest routes that you should try first but are less important than the target molecule under discussion. We have suggested that small rings may dominate strategy so our first disconnection **a** of the ketone **36** leads to the obvious addition of a carbene to the alkene **37**. Unfortunately, the alkene **37** is a tautomer of α-naphthol and cannot be made. The six-membered ring must be disconnected first **36b**.

Any pair of the three bonds in the cyclopropane ring such as **35a** could now be disconnected but none is very favourable. We should much rather use a diazocarbonyl compound such as **42** to make the carbene. That will mean chain extension after cyclopropane formation.

So, in this instance, cyclopropane formation is the first rather than the last step in the synthesis. Addition of ethyl diazoacetate **42**; R = Et to styrene **41** gave a mixture[7] of *cis* and *trans* isomers of **40**; R = Et. Only the *cis* isomer will be able to cyclise so separation[8] of the free acid gave 34% yield of *cis*-**43**. Chain extension by the Arndt–Eistert procedure[9] (chapter 31) gave *cis*-**35** and the acid chloride duly cyclised to **36**.

Regioselectivity Suggests a Change of Strategy

Raphael[10] needed the diketone **44** for his synthesis of strigol (a compound that triggers the germination of the parasitic plant witchweed). At first sight a route via electrophilic substitution and the aromatic ketone **45** looks promising. But the regioselectivity is wrong. The OH dominates and directs *ortho*. Indeed it was already known[11] that the product of reaction between **46** and **47** was isomeric **48**.

A better strategy emerges from disconnection of the six-membered ring **44a**. Aldol disconnection reveals a triketone with two 1,4-dicarbonyl relationships **49**. An ideal disconnection would correspond to a reagent for the d[1] synthon **51** that can do conjugate additions to both **50** and **52**.

This worked well with nitromethane **53** as the d[1] reagent. Conjugate addition[12] to **50** and then **52** gave the diketone **55** that cyclised to **56** in acid. Hydrolysis of the nitro group with TiCl$_3$ (chapter 22) gave the required ketone[13] **44**.

A Synthesis of the Anticancer Compound Taxol®

Taxol, the anti-cancer compound from yew trees[14] (*Taxol* spp.) has two six- and one eight-membered carbocyclic rings **58**. Nicolaou[15] saw that two C–C disconnections would give two simple six-membered ring precursors **57** and **59**.

Compound **57** looks like a Diels-Alder adduct but the ketone must be changed **60** and it is much simpler if one alkoxy group is removed **61**. Diels-Alder disconnection then gives the known dienophile **62** and a functionalised diene **63**.

Nicolaou found that the closely related diene **64** was known from some basic chemistry, already reported from the 1960s, using just aldol, Grignard reagents and elimination.[16]

This part of the synthesis went very well: the Diels-Alder reaction required a high temperature in a sealed tube but gave an excellent yield of **61**.

Hydrolysis and reacetylation gave the ketone **67**, which was protected and oxidised with SeO_2 to give the enone **68**. Reduction and deprotection gave **57**; R^1=H. The other starting material **59** was also made by a Diels-Alder reaction as described in the workbook.

A Synthesis of Sarracenin

For a remarkable strategy we need look no further than a synthesis of sarracenin **69** by Yin and three Changs.[17] The rings in **69** are all heterocyclic but the starting material for the synthesis contains only carbocyclic rings **73**. Sarracenin contains several 1,1-diX relationships and they can be disconnected in any order. So disconnection of the acetal **69** reveals an alcohol, an aldehyde and a hemiacetal **70** that can be disconnected to reveal a second aldehyde and an enol **71**. Rewriting the enol as an aldehyde we arrive at a compound with no rings **72** but lots of stereochemistry, better drawn as **72a**. There are several 1,5-diCO relationships here but the normal conjugate addition approach to these would make stereochemical control difficult. So it was decided to use the reconnection strategy normally used for six-membered rings (chapter 36). In fact the starting material was **73** where X is a heteroatom.

The synthesis of this starting material **73**; X = SMe also uses interesting strategy. The first step was a Diels-Alder reaction between a cyclopentadiene **77** and the dienophile **62** used by Nicolaou in his taxol® synthesis. The diene was made from the anion of cyclopentadiene **74** and ethyl formate,[18] and the enolate **76** transformed into the enol acetate **77**. Diels-Alder addition of **62** and hydrolysis of the enol ester gave the adduct[19] **78**.

The aldehyde **78** was protected as an acetal **79** and the ketone **80** revealed by hydrolysis. Methylation of the lithium enolate occurred on the less hindered bottom face **81** to set the scene for the remarkable chemistry to come.

Ring expansion by chemistry discussed in the workbook gives **82** whose enolate is sulfenylated **83** and cyclised in acid solution with hydrolysis of the acetal to give **84** and the five-membered rings in **73** start to appear.

Mesylation of the free OH sets up an ideal arrangement for a base-catalysed fragmentation **85** and suddenly the two fused five-membered rings are there. Acid moves the alkene into the more stable position and esterification gives **73**. Notice that there are three five- and one seven-membered ring in **84**: fragmentation cleaves one five- and the seven-membered ring leaving the two rings that we want. One C–C bond of one of the five-membered rings was made in the original Diels-Alder reaction while the other five-membered ring was the diene.

References

1. E. J. Corey, M. Behforouz and M. Ishiguro, *J. Am. Chem. Soc.*, 1979, **101**, 1608.
2. G. Stork, D. F. Taber and M. Marx, *Tetrahedron Lett.*, 1978, 2445; see footnote 3.
3. J. M. Conia and P. Beslin, *Bull. Soc. Chim. Fr.*, 1969, 483; R. L. Cargill, J. R. Dalton, S. O'Connor and D. G. Michels, *Tetrahedron Lett.*, 1978, 4465.
4. D. F. Taber, *J. Org. Chem.*, 1976, **41**, 2649.
5. R. L. Cargill, J. R. Dalton, S. O'Connor and D. G. Michels, *Tetrahedron Lett.*, 1978, 4465.
6. J. M. Hook and L. Mander, *J. Org. Chem.*, 1980, **45**, 1722; J. M. Hook, L. Mander and R. Urech, *J. Am. Chem. Soc.*, 1980, **102**, 6628.
7. A. Burger and W. L. Yost, *J. Am. Chem. Soc.*, 1948, **70**, 2198.
8. C. Kaiser, J. Weinstock and M. P. Olmstead, *Org. Synth.*, 1979, **50**, 94.
9. M. J. Perkins, N. B. Peynircioglu and B. V. Smith, *J. Chem. Soc., Perkin Trans. 2*, 1978, 1025.
10. G. A. MacAlpine, R. A. Raphael, A. Shaw, A. W. Taylor and H.-J. Wild, *J. Chem. Soc., Chem. Commun.*, 1974, 834.
11. B. R. Davis and I. R. N. McCormick, *J. Chem. Soc., Perkin Trans. 1*, 1979, 3001.
12. W. D. S. Bowering, V. M. Clark, R. S. Thakur and Lord Todd, *Annalen*, 1963, **669**, 106.
13. G. A. MacAlpine, R. A. Raphael, A. Shaw and A. W. Taylor, *J. Chem. Soc., Perkin Trans. 1*, 1976, 410.
14. K. C. Nicolaou and R. K. Guy, *Angew. Chem. Int. Ed.*, 1995, **34**, 2079.
15. K. C. Nicolaou, C.-K. Hwang, E. J. Sorensen and C. F. Clairbourne, *J. Chem. Soc., Chem. Commun.*, 1992, 1117.

16. M. A. Kazi, I. H. Khan and M. Y. Khan, *J. Chem. Soc.*, 1964, 1511; I. Alkonyi and D. Szabó, *Chem. Ber.*, 1967, **100**, 2773.

17. M.-Y. Chang, C.-P. Chang, W.-K. Yin and N.-C. Chang, *J. Org. Chem.*, 1997, **62**, 641.

18. K. Haffner, G. Schultz and K. Wagner, *Annalen*, 1964, **678**, 39.

19. E. D. Brown, R. Clarkson, T. J. Leeney and G. E. Robinson, *J. Chem. Soc., Perkin Trans. 1*, 1978, 1507.

38 Strategy XVII: Stereoselectivity B

Background Needed for this Chapter References to Clayden, *Organic Chemistry:*
Chapter 33: Stereoselective Reactions of Cyclic Compounds; Chapter 34: Diastereoselectivity.

This chapter follows on from chapter 12 where we introduced some basic ideas on stereocontrol. Since then we have met many stereospecific reactions such as pericyclic reactions including Diels-Alder (chapter 17), 2 + 2 photochemical cycloadditions (chapter 32), thermal (chapter 33) cycloadditions, and electrocyclic reactions (chapter 35). Then we have seen rearrangements where migration occurs with retention at the migrating group such as the Baeyer–Villiger (chapters 27 and 33), the Arndt–Eistert (chapter 31) and the pinacol (chapter 31).

We have expanded our collection of stereoselective reactions even more in the making of alkenes by the Wittig reaction (chapter 15), from acetylenes (chapter 16), by thermodynamic control in enone synthesis (chapters 18 and 19) and in sigmatropic rearrangements (chapter 35). We have seen that such *E*- or *Z*-alkenes can be transformed into three-dimensional stereochemistry by the Diels-Alder reaction (chapter 17), by electrophilic addition (chapters 23 and 30), by carbene insertion (chapter 30) and by cycloadditions to make four-membered rings (chapters 32 and 33).

With so many methods of stereochemical control now available, it is time to look in general at syntheses where stereochemistry dominates the strategy. This is a very large subject and this chapter deals only with *diastereo*-selectivity. Our *Strategy and Control* expands on diastereoselectivity and the synthesis of single enantiomers.

Synthesis of Molecules with Many Chiral Centres

The Prelog–Djerassi Lactone

At the start of the analysis when you have done no more than recognise the FGs and note special features (such as rings) or easy disconnections, note also the number of chiral centres and their relationship to each other. The Prelog–Djerassi lactone **1** is an important intermediate in the synthesis of macrolide antibiotics.[1] It has a six-membered lactone ring and a separate carboxylic acid. More to the point, it has four chiral centres **1a**. Three (1–3) are adjacent and one (5) separate. We might say that the three adjacent centres should be easy to control because they are next to each other but that we might have trouble with C-5. Another way to look at it is to say that the three round the six-membered ring (2, 3 and 5) should be easy to control, as the conformation of six-membered rings is so well understood, but that we might have trouble with C-1. The most obvious disconnection **1b**, that of the lactone, doesn't help much as **2** is now open-chain.

One appealing strategy is to set up C-5 and one of the three adjacent centres initially and then control the other two from that one. One disconnection that leads to immediate simplification **1c** uses the 1,3-diX relationship between the functional groups. Aldol disconnection of **3** leads to the nearly symmetrical **4** with a 1,4-diCO relationship. Differentiating the carbonyl groups would be easy with the cyclic anhydride **5** and this was the chosen starting material.[2]

Since **5** is symmetrical (it has a plane of symmetry and is achiral) it doesn't matter which carbonyl reacts so the sequence of symmetry breaking with ethanol, formation of the acid chloride **7** and reduction of the acid chloride gives the aldehyde **8** ready for the aldol step.

Bartlett and Adams[2] chose to use a Wittig reaction to make the alkene **3** in excellent yield and conditions were found experimentally to cyclise **3** to **1**. The product was a mixture of **1** and its C-2 epimer but these could be separated by chromatography.

There are many more syntheses of this important compound and interesting strategies include a reconnection of the 1,7-related carboxylic acids **2a** to offer a seven-membered ring compound as starting material[3] **10**.

The methyl group at C-2 was introduced by cuprate addition to the enone **11** on the face opposite the large TBDMS group and the lithium enolate **12** trapped with Me_3SiCl to give **13**. This silyl enol ether was ozonised, reduced, and the silyl group hydrolysed whereupon oxidation gave spontaneous lactonisation to **1** in 12% yield from **11**.

A lactone invites a Baeyer–Villiger rearrangement, in this case from the cyclopentanone **14**. The more highly substituted group migrates (chapter 27) but which is that? The ambiguity is easily removed by disconnecting the one methyl group that can easily be added by alkylation of an enolate. Now the side-chain carbon must migrate and does so with retention of configuration. This ketone **15** can be made in various ways.[4] A review article[5] has more details and many more syntheses.

Using the Diels-Alder Reaction

A similar stereochemical problem arises with the dialdehyde **16** needed for alkaloid synthesis. Reconnecting the 1,6-diCO relationship **17** and removing the acetal reveals an obvious Diels-Alder adduct **18** from the enone **19** and butadiene. The only substituent on the nearly flat enone **19** is the methyl group and the Diels-Alder reaction does indeed give the right diastereomer.[6]

Stereochemical Control in Folded Molecules

cis-Fused smaller rings (four-, five- and six-membered) have folded conformations rather like a half-opened book. The 4/4 fused system **20** has nearly flat rings and looks just like a book **20a**. It has two faces: the inside, concave or *endo*-face and the outside, convex or *exo*-face. Reagents much prefer to approach from the outside and it can be difficult to get substitution on the inside. So the 5/4 fused ketone **21** forms an enolate that is alkylated on the outside face. Notice that this means the new substituent is on the *same* face as the ring-junction hydrogens **22**.

The anti-cancer compound coriolin **23** has three fused five-membered rings and two epoxides. Notice that the 3/5 and both 5/5 ring fusions are *cis*. There have been many syntheses of coriolin, most using stereochemistry from folded precursors.[7] We shall feature a couple of examples. Matsomoto's synthesis involves the hydroboration of alkene **24**. The addition of borane is *cis* **25** and the boron is replaced by OH with retention of configuration to give **26**. The hydroboration occurred on the outside of the molecule, on the same face as the ring junction hydrogens.[8]

Ikegami's synthesis has two steps of particular interest. Allylation of the enolate of **27** puts the new allyl group on the outside and forces the old methyl group onto the inside **28**. The planar enolate is an intermediate. The final step puts in both epoxides on the outside of the molecule from the dienone **29** mainly through complexation of VO^+ with one of the free OH groups.[9] This also makes both the right-hand 5/5 and the 5/3 *cis* fused.[10]

We discussed the synthesis of tricyclic **30** in chapter 32 (compound **12** there) but did not discuss how the stereochemistry of the ring junctions was proved. Rings A/B form a folded system so, when the ketone is reduced with $NaBH_4$, we expect the reagent to approach from the *exo*-face, on the same side as the ring junction H atom. In fact the lactone **32** was isolated.[11] This compound could be made only if the OH was on the same side of the molecule as the CO_2Me group so **31** must have the structure predicted and **30** must have the structure shown.

30 **31** **32**

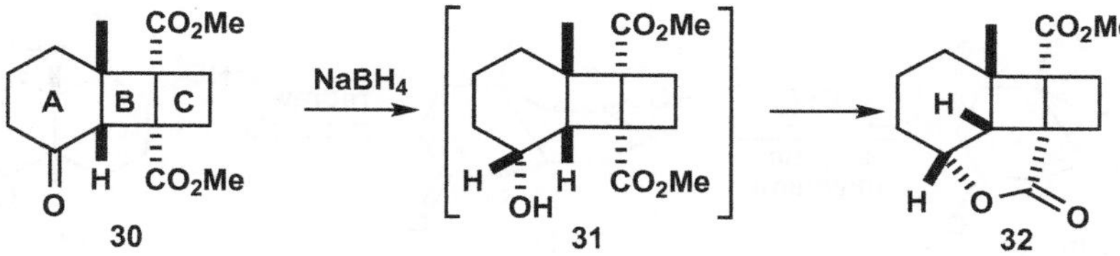

When we come to two six-membered rings fused together **33**, it is not so obvious that there is an inside and an outside, but if both rings have the chair conformation **33a** there is an inside and an outside. Even if there is only one ring-junction substituent **34** there is still a distinction.

33 **33a** **34**

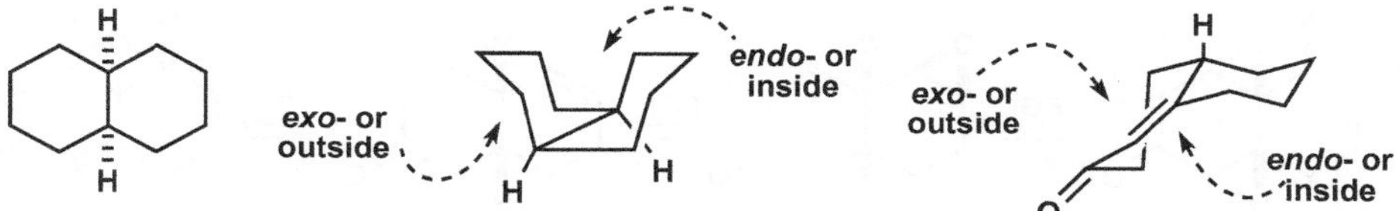

A clear example is the catalytic reduction of the Robinson annelation product **36**. Hydrogen adds to the top face of the alkene to give the *cis*-fused system **35**. The point is that the alkene must sit on the catalyst surface and that is much easier on the top face (cf. **34**). The axial methyl group at the ring junction is small compared with the other ring. But if the reagent approaches on the other ring next to the axial methyl group, as in borohydride reduction to give **37**, it can attack from the bottom face.

35 **36** **37**

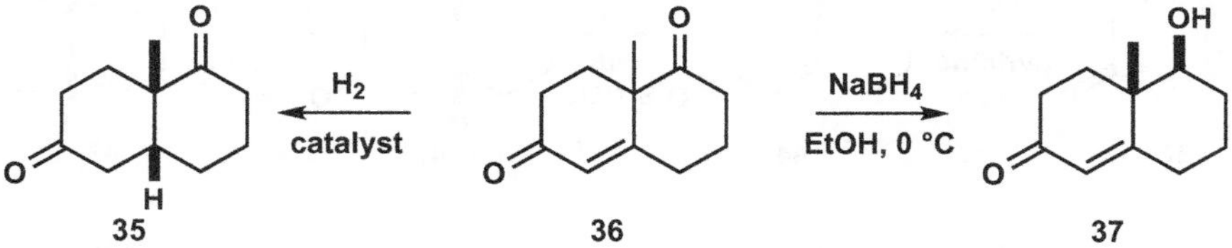

A Synthesis of Copaene

The strange terpene copaene **38** has two six-membered rings with a four-membered ring trapped between them. Heathcock[12] chose to put in some functionality **39** to help disconnections. Approaches based on 2 + 2 photochemical cycloadditions are unlikely to help: starting materials will be 10-membered rings with two *trans* alkenes **40**.

38; α-copaene **39** **40**

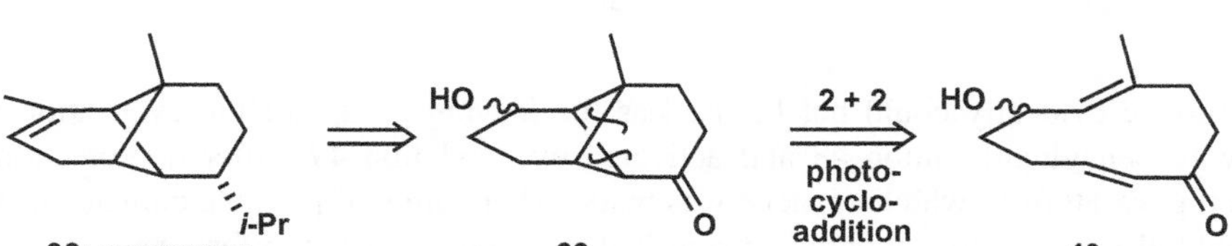

A disconnection of just one of the bonds in the four-membered ring might use the enolate of the ketone to displace a leaving group X **41** and we have a much more helpful *cis*-decalin **41a**.

With X and OH next to one another, an epoxide **42** looks a good bet. This would have to come from an alkene **43** and there must be a possibility of getting this from the Robinson annelation product **36** whose reduction we have just been discussing.

Both reductions (of C=O and C=C in **36**) are needed and **37** was chosen as the starting material. Tosylation provided the leaving group **44** and the hydrogenation gave the expected selectivity. The equatorial tosyl group is more-or-less in the plane of the alkene and has little effect. Elimination can go only one way and we have **43** in a few steps.

The ketone had to be protected before epoxidation (avoiding the Baeyer–Villiger reaction?) and epoxidation then occurred on the outside of the folded molecule **47**.

The epoxide evidently could not be used as the leaving group for the cyclisation so it was opened with benzyloxide anion **48** and activated by tosylation **49**. After deprotection, enolate cyclisation gave **50** from which copaene was made. There are two possible enolates from **49**: the other would also give a four-membered ring if it cyclised but it is too far away.

The oxyanion must open the epoxide to give a *trans* diaxial product **48a** by attacking the less hindered end of the epoxide. This is inevitably from the *endo* face but attack at the other end of the epoxide would have to be from right inside the fold. Being a *cis* decalin, the product can equilibrate to the equatorial conformer **48b** and the arrangement for cyclisation is perfect **51**.

Summary of Stereoselectivity on Folded Molecules

Reagents, whether electrophilic or nucleophilic, prefer to approach the 'outside' or '*exo*' face of folded molecules. These are *cis*-fused four-, five-, or six-membered rings. If a substituent is needed on the 'inside' or '*endo*' face, it must be added first, that is disconnected second **54**, as in the synthesis of the ketone **54**. If there is only one substituent and it is *endo*, R^2 can be H.

The Synthesis of Juvabione

Juvabione **55** is a juvenile hormone mimic produced by the balsam fir as a defence against bugs. It prevents the larvae metamorphosing into adults. It has a six-membered ring and two adjacent chiral centres, one on the ring and one outside. The only obvious disconnection is of the unsaturated ester that gives a tricarbonyl compound **56** with a 1,6-diCO relationship. Reconnection leads back to juvabione but the other gives a different cyclohexene **57**. No doubt this compound could be made but it does not suggest any way of controlling the stereochemistry.

A more drastic series of disconnections removes the isobutyl side chain **55a** and then replaces the unsaturated ester with a ketone that appears to be on the 'wrong' carbon atom.[13]

The reason for this becomes plain when you see that the new 1,6-diCO relationship does allow a very interesting reconnection. Adjustment of the oxidation state allows a reconnection to a lactone **61** that should be the product of a Baeyer–Villiger rearrangement on the ketone **62**.

Now all three chiral centres are in a rigid framework **62a** making it easier to control them. An alkene was introduced and you might think Schultz and Dittani were going to do an aldol disconnection **63** next. The diketone **64** is symmetrical and might well cyclise to **63**.

However, that was not how they made **63**. They preferred a method we met in chapter 37 where a cyclohexenone was used in alkylation. Here there will be *three* alkylations of the enol ether **68**, one with a nucleophile and two with electrophiles. So the disconnections cleave a C_3 side-chain in two alkylations **63a** and **66** with a nucleophilic addition of a methyl group in between.

The first few steps used **67**; X = I, Y = Cl to ensure faster reaction at one end. The intermediate **66**; X = Cl did not cyclise in good yield, but replacing Cl with the better leaving group I ensured an excellent cyclisation. Note that although we said in chapter 37 that such enol ethers had to be symmetrical, this alkylation of **68**, followed by addition of MeMgBr and rearrangement gives only one isomer of **65**.

68; R = Et　　　　　**66; X = Cl**　　　　　**65; 50% yield from 68**

The hydrogenation of **63** occurred quantitatively and with high stereoselectivity (25:1 in favour of **62**). The regioselectivity was similarly good in the Baeyer–Villiger rearrangement (12:1) in favour of **61** over the isomeric lactone. Attempts to convert the lactone directly to **60**; X = *i*-Bu gave only low yields.

63　　　　　**62; 100% yield**　　　　　**61; 58% yield**

Cleavage of the lactone with methanol and protection of the OH group with the bulky *t*-BuMe₂Si group gave **69** which did react cleanly with *i*-BuMgCl to give the ketone **70** in 90% yield from **61**. Now the transformation of **70** into juvabione follows Ficini's synthesis.[14]

60; X = OMe　　　　　**69: R = TBDMS**　　　　　**70: R = TBDMS**

Protection of the free ketone **70** as the acetal and desilylation with fluoride give the alcohol **71**, which was oxidised to the ketone **72** with Cr(VI) in the form of pyridinium chlorochromate (PCC).

70: R = TBDMS　　　　　**71; 91% yield**　　　　　**72; 97% yield**

The new ketone can be used to add the ester group regioselectively to the less hindered side (though this is not important) and the ketone is reduced to a mixture of epimers of the alcohol **73**. Tosylation and elimination give the conjugated alkene by the E1cB mechanism so that the stereochemistry of the OTs group is irrelevant. Deprotection gives juvabione **65**.

72　　　　　**73; 60% yield**　　　　　**55; 55% yield**

You may well think that this is a long synthesis but other shorter ones fail to control the stereochemistry. The shortest[15] gets quickly to the enone **75** by Diels-Alder and HWE reactions but the stereoselectivity of the reduction is only moderate.

Birch's synthesis falls down at the first step, a Diels-Alder reaction between the diene **76** and an enone gives the whole skeleton of juvabione but as a 1:1 mixture of *exo* **77** and *endo* **78** isomers. It is also a long way from these compounds to the final product.[16]

Stereochemistry is both the most difficult and the most interesting aspect of the design of organic syntheses. In recent years great advances have been made in diastereoselectivity[17] and enantioselectivity.[18] These themes are pursued in *Strategy and Control*.[19]

References

1. S. F. Martin and D. E. Guinn, *Synthesis*, 1991, 245.
2. P. D. Bartlett and J. L. Adams, *J. Am. Chem. Soc.*, 1980, **102**, 337.
3. J. D. White and Y. Fukuyama, *J. Am. Chem. Soc.*, 1979, **101**, 226.
4. P. Grieco, Y. Ohfune, Y. Yokoyama and W. Owens, *J. Am. Chem. Soc.*, 1979, **101**, 4749; P. M. Wovkulich and M. R. Uskokovic, *J. Org. Chem.*, 1982, **47**, 1600.
5. S. F. Martin and D. E. Guinn, *Synthesis*, 1991, 245.
6. T. Harayama, M. Takanati and Y. Inubushi, *Tetrahedron Lett.*, 1979, 4307.
7. J. Mulzer, H.-J. Altenbach, M. Braun, K. Krohn and H. U. Reissig, *Organic Synthesis Highlights*, VCH, Weinheim, 1991, page 323.
8. T. Ito, N. Tomiyoshi, K. Nakamura, S. Azuma, M. Izawa, M. Muruyama, H. Shirahama and T. Matsumoto, *Tetrahedron Lett.*, 1982, **23**, 1721; *Tetrahedron*, 1984, **40**, 241.
9. S. Danishefsky, R. Zamboni, M. Kahn and S. J. Etheredge, *J. Am. Chem. Soc.*, 1980, **102**, 2097 and 1981, **103**, 3460.
10. M. Shibasaki, K. Iseki and S. Ikegama, *Tetrahedron Lett.*, 1980, **21**, 3587; K. Iseki, M. Yamazaki, M. Shibasaki and S. Ikegama, *Tetrahedron*, 1981, **37**, 4411.
11. G. L. Lange, M.-A. Huggins and E. Neiderdt, *Tetrahedron Lett.*, 1976, 4409.
12. C. H. Heathcock, R. A. Badger and J. W. Patterson, *J. Am. Chem. Soc.*, 1967, **89**, 4133.
13. A. G. Schultz and J. P. Dittami, *J. Org. Chem.*, 1984, **49**, 2615.
14. J. Ficini, J. d'Angelo and J. Nioré, *J. Am. Chem. Soc.*, 1974, **96**, 1213.

15. M. Fujii, T. Aida, M. Yoshikara and A. Ohno, *Bull. Chem. Soc. Jpn.*, 1990, **63**, 1255.
16. A. J. Birch, P. L. Macdonald and V. H. Powell, *Tetrahedron Lett.*, 1969, 351.
17. Clayden, *Organic Chemistry*, chapter 34.
18. Clayden, *Organic Chemistry*, chapter 45.
19. *Strategy and Control*, chapters 20–31.

39 Aromatic Heterocycles

Background Needed for this Chapter Reference to Clayden, *Organic Chemistry:* Chapters 43 and 44: Aromatic Heterocycles, Structures, Reactions and Synthesis.

Aromatic heterocycles come in many shapes and forms. They may be five-membered rings with one or two heteroatoms such as furan **1**, imidazole **2** or thiazole **3**. They may be six-membered rings such as pyridine **4** or pyrimidine **5** or even have two rings fused together such as indole **6** or isoquinoline **7**. Substituted versions are important in pharmaceuticals, agrochemicals, perfumery, food and colour chemistry. Encyclopaedias and books[1] have been written on this vast subject as more than half of all known organic compounds are aromatic heterocycles. This chapter is a brief introduction to the general strategies available for the synthesis of aromatic heterocycles.

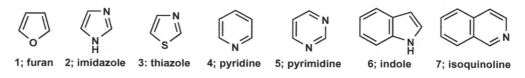

1; furan   2; imidazole   3: thiazole   4; pyridine   5; pyrimidine   6; indole   7; isoquinoline

Carbon–Heteroatom Disconnections

Many heterocyclic rings are made by the formation of a carbon–heteroatom bond and it is important when planning this to get the oxidation level of the carbon electrophile right. If we disconnected either C–N bond of the pyrrole **8**, we get back to a ketone and an amine **9**. If we disconnected the imine in imidazole **10** we should also get back to a ketone and an amine **11**. Both **9** and **11** are unstable and neither is likely to be a real intermediate but the point is worth making. These carbon electrophiles are at the aldehyde or ketone oxidation level.

By contrast, disconnection of the pyridone **12** gives a similarly unstable amine and a carboxylic acid **13**. We are never going to make **8**, **10** or **12** from these intermediates but it will

Organic Synthesis: The Disconnection Approach. Second Edition   Stuart Warren and Paul Wyatt
© 2008 John Wiley & Sons, Ltd

be important to recognise that the marked carbon atoms in **12a**, **14** and **15** are at the carboxylic acid oxidation level.

We can put this into practice immediately with the synthesis of pyrroles **16**. Disconnection of *both* C–N bonds gives a very reasonable intermediate, the 1,4-diketone **17**. This will need to be made by one of the methods from chapter 25 and treatment with ammonia gives **16**. On the other hand, if the furan **18** is needed, no heteroatom needs to be added and treatment with acid cyclises the diketone **17** to the furan **18**.

Synthesis of Pyrroles

A simple example is the pyrrole **19** required for the synthesis of the anti-inflammatory clopirac. Disconnection of the two C–N bonds reveals the diketone **20**, available as 'acetonylacetone', and the simple aromatic amine **21**. The synthesis is to mix the two together. This synthesis makes *N*-substituted pyrroles available.[2]

If the 1,4-dicarbonyl compound is unsymmetrical, it will have to be made by methods such as those described in chapter 25. Examples such as **22** appear in papers by Yadav.[3] The 1,4-diketone **23** can be disconnected at a branchpoint with the idea of using a d^1 reagent for BuCHO in conjugate addition to the enone **24**.

Any of the d^1 reagents suggested in chapter 25 would do but they preferred a catalytic method using the thiazolium salts **25** devised by Stetter.[4] These are also aromatic heterocyclic compounds and their mechanism of action is discussed in the workbook. They chose microwaves to assist the first step and the unusual Lewis acid catalyst bismuth triflate for the second in the 'ionic liquid' bmim[5] **26**, also an aromatic heterocyclic compound. These dialkyl imidazolium salts are easily made by double alkylation of imidazole.[6]

Thiazoles

When there are two different heteroatoms in a five-membered aromatic ring, questions of regio-selectivity often arise. The unsymmetrical thiazole **27** might be disconnected at the imine to give the unstable primary enamine **28** and then at the thiolester to give an acylating agent and the undoubtedly very unstable **30**. Note that **30** must have SH and NH$_2$ on the same side of the alkene for cyclisation to be possible. We want to find a better way.

But the more heteroatoms, the more alternatives. We could disconnect the enamine first **27a** and the C–S bond second **31**. This suggests a reasonable α-halo-ketone **33** and an unstable-looking imine **32**. Fortunately this is just a tautomer of the thioamide **34**. Though thioketones are unstable, thio-amides are stable thanks to extra conjugation.

This is the strategy followed in most thiazole syntheses. The regioselectivity issue is which way the reagents combine. There are two possibilities: the sulfur could attack either the ketone or the saturated carbon atom as can the nitrogen. But sulfur is excellent at S$_N$2 reactions while nitrogen is better at addition to carbonyls. So **27** and not **35** is the product. No intermediates are isolated: once either the C–S or the C–N bond is formed, cyclisation and aromatisation are fast. This means that aromatic heterocycles are easier to make than the non-aromatic ones.

A simple example is the anti-inflammatory fentiazac[7] **36**. Doing both disconnections at once we get available thiobenzamide **37** and the α-halo-ketone **38**. This can be made from the parent ketone **39**, available by a Friedel-Crafts reaction using a cyclic anhydride (chapter 25).

36; fentiazac   **37**   **38**   **39**

The synthesis of the Stetter thiazolium salt **25** uses a different strategy. The first disconnection is obviously of the benzyl group and then we need the α-chloro-ketone **41** for reaction with thio-formamide.

25   **40**   **41**   **42**

You might think that **41** could easily be made by chlorination of the hydroxy-ketone **43**, but how are we to control enolisation? One way would be to add a controlling CO_2Et group **44** as we can then do a C–C disconnection back to ethylene oxide and ethyl acetoacetate **45**.

41   **43**   **44**   **45**

We have already met this sequence in chapter 25: the intermediate is in fact the lactone **46** that is chlorinated in excellent yield and gives the thiazole in a few steps.[8]

46   **47; 84% yield**

The thiazole **40** is available as it is an intermediate in the manufacture of vitamin B_1 and one patented synthesis uses the dichloro-compound **49** to make **41** by rather a different route.[9]

48   **49**   **41**   **40**

Six-Membered Rings: Pyridines

Disconnection of both C–N bonds of a pyridine **50** gives an ene-dione **51** but the alkene has to be *cis* for cyclisation to be possible and conjugated *cis*-enones are rather unstable. It is usually easier to remove the double bond to reveal the saturated 1,5-diketone **52** that can be made by the methods of chapter 21. This usually means conjugate addition of an enolate to an enone.

Treatment of the diketone **52** with ammonia gives the dihydropyridine **53** that is very easily oxidised by a variety of oxidants to the pyridine **50** itself. A hydrogen from C-4 is very easily removed as the product is aromatic. If you know some biological chemistry you will recognise a similarity to NADPH.

If you don't want to be bothered with the oxidation, you can use hydroxylamine instead of ammonia. The intermediate is now unstable and eliminates water **54** very easily. One of the two marked Hs at C-4 is lost as a proton with cleavage of the weak N–O bond to give the pyridine **50** and water.

A simple example shows just how easy this is. The bicyclic pyridine **55** gives the diketone **56** by disconnection and FGA. Disconnecting **56** at the branchpoint suggests some enolate equivalent of cyclohexanone and the enone **57**.

As explained in chapter 21, vinyl ketones such as **57** are unstable and we often prefer to use the Mannich base instead. This example works spectacularly well. Heating the Mannich base **58** with cyclohexanone gives the 1,5-diketone **55** that combines with hydroxylamine to give the pyridine **54** both reactions giving excellent yield.[10]

Pyrimidones exist as the amide tautomer **60** rather than the enol tautomer **59** unlike the case with phenols **61**. The enol tautomer of a phenol **61** is aromatic but the keto tautomer **62** is not.

Both tautomers of pyridones are aromatic as the amide nitrogen in **60** has a delocalised pair of electrons and the extra stabilisation of the carbonyl group carries the day.

When drawing disconnections it doesn't matter which tautomer you use. If we disconnect both C–N bonds **63** we get a keto-acid **64** and removal of the alkene gives a simple 1,5-dicarbonyl compound **65**. Reaction with ammonia followed by oxidation gives the pyrimidone.

Pyrimidines

Pyrimidines are pyridines with two nitrogen atoms having a 1,3-relationship **66**. They occur in nucleic acids as pyrimidine bases such as cytosine **67** and thymine **68** and you will notice that these are pyrimidones. An important new type of anti-cancer drug Glivec **69** has a pyrimidine core with a linked pyridine ring.

The compound Aphox, that kills greenfly without harming ladybirds, is a pyrimidine **70**. Disconnection of the ester side chain reveals a pyrimidine **71** that we should rather draw as a pyrimidone **72**. Disconnection of two C–N bonds gives simple starting materials: available dimethyl guanidine **73** and the acetoacetate derivative **74**.

The synthesis[11] used ethyl acetoacetate which was methylated and cyclised with the guanidine **73** to give the pyrimidone **72**: acylation on oxygen gives Aphox directly.

Benzene-Fused Heterocycles: Indoles

The most important of the heterocycles fused to benzene rings are the indoles **75**. The obvious enamine disconnection gives **76** which would certainly cyclise to the indole, but how are we to make **76**? As a result of this difficulty, many special reactions have been invented to make indoles and the most important is the Fischer indole synthesis.[12] A phenylhydrazone **77** of a ketone or aldehyde is treated with acid or Lewis acid and the product is an indole.

In essence, the phenylhydrazone **77**, formed from the ketone **78** and $PhNHNH_2$, tautomerises to an enamine that can undergo a [3,3]-sigmatropic rearrangement, with cleavage of the weak N–N bond **79**, to give an unstable intermediate **80** that aromatises to **81**. Cyclisation of the NH_2 group onto the imine and loss of ammonia gives the indole.

The easiest way to work out how to make an indole is to disconnect the only two bonds formed in the reaction: the C–C and C–N bonds in **82** and write the hydrazine **83** and ketone **78** required. There are two questions of selectivity: here we ask which of the two *ortho* positions (black blobs in **84**) will react? Answer: it doesn't matter—they are the same.

The other question is: which side of the ketone will 'enolise' or more accurately form the enamine **79**? Here it may be easy as they may again both be the same as in **86**. In diagrams **84** and **86**, the dotted line shows the symmetry.

Example: The Synthesis of Indomethacin

Indomethacin **87** is a Merck non-steroidal anti-inflammatory. Amide disconnection reveals a simpler indole **88** and the Fischer disconnection gives the two starting materials **89** and **90**.

It should be obvious that the hydrazine has the ideal symmetry—both *ortho*-positions are the same—but the ketone does not.

The hydrazine **89** is made from the amine by nitrosation and reduction and the keto-acid **90** is available as levulinic acid. Now comes the big question: when the Fischer indole synthesis is carried out on the hydrazone **91**: which enamine is formed, the one we want **92**, or the one we don't want **93**? Since the Fischer indole is an acid- (or Lewis acid-) catalysed reaction we expect the more substituted enamine **92** to be favoured.

And it is. The *t*-butyl ester of **90** was used and a good yield of the indole **96** was obtained just by heating in ethanol with HCl as catalyst. The acylation required the acid chloride **97** and the *t*-butyl ester was 'hydrolysed' by heating.[13]

Making Bonds to pre-Formed Heterocycles

So far we have concentrated on making complete heterocycles with substituents. So indomethacin **87** was made with one substituent (OMe) on the benzene ring and two on the pyrrole ring. Only the substituent on nitrogen was added after the indole was formed. Now we shall consider what reactions can be used to add substituents to heterocycles after they are formed. This will usually be by electrophilic or nucleophilic aromatic substitution. The most important distinction between five- and six-membered rings is that pyrroles, indoles and furans are good at *electro*philic whilst pyridines and pyrimidines are good at *nucleo*philic substitution.

Electrophilic Substitution on Pyrroles, Indoles and Furans

These five-membered rings have lone pair(s) delocalised from the heteroatom round the ring and are 'electron-rich'. They react all too easily with electrophiles and are unstable in acid whether protic or Lewis. We have to find reactions that can be used in neutral or only weakly acidic solution. The synthesis of tolmetin **99** illustrates the two most important reactions.[14] The disconnection of the ketone would lead naturally to an $AlCl_3$-catalysed Friedel-Crafts reaction between the acid chloride **100** and the pyrrole **101**.

But this mixture would decompose the pyrrole **101**. We use instead the Vilsmeier acylation, replacing the acid chloride by the tertiary amide and $AlCl_3$ by $POCl_3$. The amide is very unreactive but combines with $POCl_3$ **102** to give a reactive species that does attack the pyrrole **104** in the right position to give, after rearomatisation **105**, the iminium salt **106** that is hydrolysed to the ketone **101**. You will notice the similarities to electrophilic substitution on benzene.

We still have to make the pyrrole with the alkyl side chain for this acylation reaction. Friedel-Crafts alkylation is not an option but pyrroles are reactive enough to do the Mannich reaction. Formaldehyde and an amine combine to give another iminium salt **107** that reacts with *N*-methyl pyrrole to give, after rearomatisation **109** the substituted pyrrole **110**.

You will also notice a problem: the amine **110** is not what we want. However, the tertiary amino group in Mannich products is often replaced by other functionalities: in chapter 20 we saw alkylation and elimination used to make enones. Here alkylation and substitution is used to make a nitrile **111** and it was this compound that was used in the acylation sequence. At the end, hydrolysis of the nitrile **112** gave tolmetin **99**.

Notice that both substitutions **104** and **108** occurred next to nitrogen. Indoles on the other hand react very selectively at C-3 **113** with electrophiles such as Vilsmeier and Mannich salts.[15] This is probably because reaction at C-2 **116** would disrupt the benzene ring as well as the pyrrole ring. As you have seen, substituents at C-2 are easy to put in during indole synthesis, so this is no great handicap.

Nucleophilic Substitution on Pyridines and Pyrimidines

Pyridines on the other hand are very bad at electrophilic substitution, so much so that it is hardly attempted, but they are excellent at nucleophilic substitution. A most important example is the transformation of pyridone into 2-chloro pyridine **119** and hence into any derivative needed, such as the amines **120**. The reagent is again POCl$_3$ which attacks oxygen **117** to give a very reactive intermediate with a good leaving group that is attacked by chloride ion **118**. All these reactions occur by the addition–elimination mechanism like **118**. It is essential to have at least one nitrogen in the ring for this to work but two nitrogens, as in a pyrimidine, are better.

The Synthesis of an Anti-Cancer Compound

We conclude this chapter with the synthesis of Novartis PKI 166 **121**, a new anti-cancer drug of great promise.[16] It has fused pyrrole and pyrimidine rings and the reaction we have just discussed allows disconnection of the amine from the pyrimidine ring **121**. Now we can use standard C–N disconnections on the two rings in turn **122** and **123** to reveal a much simpler starting material **124**.

The keto-acid has a 1,4-diCO relationship **124a** and the most promising disconnection **124b** gives an α-halo-ketone **125** and a curious double enamine **126**. The reaction will require selectivity as the nucleophilic carbon must displace bromide by an S_N2 reaction while the nitrogen must attack the carbonyl group. This is the 'right' way round mechanistically too.

It turns out that we must protect the phenol as its methyl ether **127** and that **126** is best used as an amidine-ester rather than the double enamine. The synthesis is then quite short. We have barely scratched the surface of aromatic heterocyclic synthesis in this chapter but the encouraging message is that cyclisation is easy and that cyclisations to form aromatic compounds are the easiest of all. Disconnect with confidence!

References

1. J. A. Joule and K. Mills, *Heterocyclic Chemistry*, Blackwell, Oxford, Fourth Edition 2000; Clayden, *Organic Chemistry*, chapters 43 and 44.

2. G. Lumbelin, J. Roba, C. Gillet and N. P. Buu-Hoi, *Ger. Pat.*, 2,261,965, 1973; *Chem. Abstr.*, 1973, **79**, 78604a.

3. J. S. Yadav, R. V. S. Reddy, B. Eeshwaraiah and M. K. Gupta, *Tetrahedron Lett.*, 2004, **45**, 5873; J. S. Yadav, K. Anuradha, R. V. S. Reddy and B. Eeshwaraiah, *Tetrahedron Lett.*, 2003, **44**, 8959.

4. H. Stetter and H. Kuhlmann, *Org. React.*, 1991, **40**, 407.

5. T. Welton, *Chem. Rev.*, 1999, **99**, 2071.

6. J. S. Wilkes, J. A. Levisky, R. A. Wilson and C. L. Hussey, *Inorg. Chem.*, 1982, **21**, 1263; P. J. Dyson, M. C. Grossel, N. Srinavasan, T. Vine, T. Welton, D. J. Williams, A. J. B. White and T. Zigras, *J. Chem. Soc., Dalton Trans.*, 1997, 3465.

7. K. Brown, D. P. Cater, J. F. Cavalla, D. Green, R. A. Newberry and A. B. Wilson, *J. Med. Chem.*, 1974, **17**, 1177.

8. I. A. Bubstov and B. Shapira, *Chem. Abstr.*, 1970, **73**, 56015.

9. T. E. Londergan and W. R. Schmitz, *U. S. Pat.*, 2,654,760, 1953, *Chem. Abstr.*, 1954, **48**, 12810a.

10. N. S. Gill, K. B. James, F. Lions and K. T. Potts, *J. Am. Chem. Soc.*, 1952, **74**, 4923.

11. F. L. C. Baranyovits and R. Ghosh, *Chem. Ind. (London)*, 1969, 1018.

12. B. Robinson, *Chem. Rev.*, 1963, **63**, 373; 1969, **69**, 227; J. A. Joule in *Science of Synthesis*, ed. E. J. Thomas, 2000, Thieme, Stuttgart, vol. **10**, page 361.

13. T. Y. Shen, R. L. Ellis, T. B. Windholz, A. R. Matzuk, A. Rosegay, S. Lucas, B. E. Witzel, C. H. Stammer, A. N. Wilson, F. W. Holly, J. D. Willet, L. H. Sarett, W. J. Holz, E. A. Risaly, G. W. Nuss and M. E. Freed, *J. Am. Chem. Soc.*, 1963, **85**, 488.

14. E. Shaw, *J. Am. Chem. Soc.*, 1955, **77**, 4319.

15. Joule and Mills, chapter 10.

16. G. Bold, K.-H. Altmann, J. Frei, M. Lang, P. W. Manley, P. Traxler, B. Wietfeld, J. Brüggen, E. Buchdunger, R. Cozens, S. Ferrari, P. Furet, F. Hofmann, G. Martiny-Baron, J. Mestan, J. Rösel, M. Sills, D. Stover, F. Acemoglu, E. Boss, R. Emmenegger, L. Lässer, E. Masso, R. Roth, C. Schlachter, W. Vetterli, D. Wyss and J. M. Wood, *J. Med. Chem.*, 2000, **43**, 183, 2310.

40 General Strategy D: Advanced Strategy

Some guidelines on strategy are collected in this final chapter and applied to a range of the types of molecules we have been discussing.

A Synthesis of Pyrazoles

The disconnection of pyrazoles **1** by the methods of the last chapter is straightforward and leads to hydrazine **2** in combination with a 1,3-dicarbonyl compound **3**. This is simply disconnected by the methods of chapter 19 to an enol(ate) of **4** and an acylating agent **5**.

So what is new here? We can save time, materials and effort if we combine two reactions in one operation. These *tandem* processes, as they are called, avoid the isolation of potentially difficult intermediates and may avoid the need for control over reactions: in chapter 19 we discussed the need for control in the acylation of enolates. Workers at Merck[1] combined the difficult acylation of enolates **7** by acid chlorides with the capture of the intermediates **9** by hydrazine to give stable pyrroles **10**. This is a summary of their method:

Organic Synthesis: The Disconnection Approach. Second Edition   Stuart Warren and Paul Wyatt
© 2008 John Wiley & Sons, Ltd

When compounds such as pyrazoles are being made to develop new drugs, a number of related compounds can be prepared at once by diversity-oriented synthesis, that is, methods designed to be general for a wide range of compounds. An example where both components are aliphatic **13** and a tri-substituted example **15** make the point. In this work, the synthesis of 25 different di- or tri-substituted pyrazoles was attempted by the same method: only one example failed.

Convergence

Long linear sequences of reactions give low yields; a linear 10-step sequence (i) with each step giving 90% yield gives only just under 35% overall. And how often does each step give 90% yield? Convergent or branching strategies make things better by reducing the longest linear sequence. Even one branch reduces the loss: sequence (ii) has only eight steps in the longest linear sequence and the yield rises to 43%.

$$A \longrightarrow B \longrightarrow C \longrightarrow D \longrightarrow E \longrightarrow F \longrightarrow G \longrightarrow H \longrightarrow I \longrightarrow J \longrightarrow TM \qquad \text{(i)}$$

$$\left.\begin{array}{l} A \longrightarrow B \longrightarrow C \\ D \longrightarrow E \longrightarrow F \end{array}\right\} \longrightarrow G \longrightarrow H \longrightarrow I \longrightarrow J \longrightarrow K \longrightarrow TM \qquad \text{(ii)}$$

But we can do much better: branching later (iii) means only five steps in the longest linear sequence and a yield of 59% while more branches (iv) and (v) give different levels of improvement. There is no magic about this: we are just 'thwarting the arithmetical demon' by our strategy and the bigger the molecule, the easier it is to devise a convergent strategy. In fact, this is what we have been doing by choosing disconnections in the middle of the molecule and at branchpoints.

$$\left.\begin{array}{l} A \longrightarrow B \longrightarrow C \longrightarrow D \longrightarrow E \\ F \longrightarrow G \longrightarrow H \longrightarrow I \longrightarrow J \end{array}\right\} K \longrightarrow TM \qquad \text{(iii)}$$

$$\left.\begin{array}{l} A \longrightarrow B \longrightarrow C \\ \qquad D \longrightarrow E \\ H \longrightarrow I \longrightarrow J \longrightarrow K \end{array}\right\} \longrightarrow F \longrightarrow G \right\} L \longrightarrow M \longrightarrow TM \qquad \text{(iv)}$$

$$\left.\begin{array}{l} A \longrightarrow B \longrightarrow C \\ D \longrightarrow E \longrightarrow F \\ H \longrightarrow I \longrightarrow J \longrightarrow K \end{array}\right\} \longrightarrow G \right\} L \longrightarrow M \longrightarrow TM \qquad \text{(v)}$$

Synthesis of Methoxatin

Methoxatin **16** is a coenzyme that allows some bacteria to use methanol as their source of carbon. It oxidises methanol by using the *ortho*-quinone on the middle ring and so it seemed reasonable to synthesise it from the much more stable benzene **17** as biological reagents must be reversible. Disconnecting either the pyrrole or the pyridine ring would lead to a linear strategy so Hendrickson[2] decided to disconnect the central benzene ring. This sounds tricky but they realised that the alkene **18** would cyclise easily to **17**.

A Wittig disconnection **18a** split the molecule into two and the decision to put the aldehyde on the pyrrole was influenced by a known synthesis from available starting materials and the hope that the phosphonium salt **20** could be made from the known pyridine **21**.

The synthesis of **23** was known to be amazingly simple: pyruvic acid **22** is mixed with ammonia! The yield is low, but who minds? If you must have a low-yielding step, it is a good idea to have it at the start of the synthesis to avoid the waste of materials and energy. In this case, so much is achieved that a low yield is acceptable. Esterification gave the diester **21**; R = Me which could be brominated with NBS (chapter 24) and combined with Ph$_3$P to give the phosphonium salt. This is the first branch complete.

The pyrrole **19** was made by a Friedel-Crafts reaction on the known and deactivated pyrrole **25**. The CO$_2$Et group deactivates C-3 and C-5 so reaction occurs at the only unaffected position. The electron-withdrawing CO$_2$Et group also makes the pyrrole less susceptible to Lewis acid degradation (chapter 39). The Wittig reaction with the ylid from **20** went in excellent yield but the product was >95% *trans* and this compound cannot cyclise.

The ingenious solution was to isomerise the alkene and cyclise it in a single operation using light to catalyse both reactions. Cyclisation of *Z*-**18** should give *trans*-**27** by a conrotatory electrocyclic reaction but the reaction was conducted with diphenyldiselenide PhSe–SePh which oxidised it to the benzene **28** in the reaction mixture. So *three* steps were combined in one.

Two linear syntheses appeared about the same time: both start with the central benzene ring and work outwards, first in one direction and then in the other. Both mark the positions on the benzene ring where oxidation will be needed. The Corey synthesis[3] puts one OMe group on the ring **29** and disconnects the pyridine ring to an indole **30** and an unsaturated ketoester **31** that might be made from available ketoglutarate **32**.

Fischer disconnection of the indole **30a** (chapter 39) gives methyl pyruvate and an aryl hydrazine that would have to be made from the corresponding diazonium salt that would have to be protected on the other amino group **36**. They decided to take a short cut by using the Japp–Klingemann reaction on the same diazonium salt with the easily made acetoacetate **35**.

The enol of **35** adds to the diazonium salt **36** and is deacylated in KOH to give the hydrazone **38** that isomerises to the enamine **39** for the Fischer indole to give the indole (*N*-formyl **30**) that is finally hydrolysed to **30** in aqueous HCl. The remarkably high regioselectivity in the cyclisation may be due to steric hindrance: it is difficult to insert a substituent between two others.

The ketodiester **31** was made from **32** by bromination and elimination. Reaction with **30** gave first the heterocycle **40** that was dehydrated and aromatised in dry HCl. Oxidation of this methoxy-compound with Ce(IV) was quite easy and methoxatin was synthesised.

Weinreb's synthesis[4] has two OMe groups where the *ortho*-quinone will be and puts in the pyrrole last. The idea was to use a Reissert indole synthesis where reduction of the nitro group in **42** would lead to condensation with the ketone.

The Reissert depends on using the nitro group to stabilise an anion on the methyl group of **43** suggesting the C–C disconnection shown on **42**. Direct nitration of **44** is expected to put the nitro group in the right place as it is the only free position on the activated (two OMe groups) benzene ring. Nitration is not expected on the pyridine ring (chapter 39) particularly as it has two CO_2Me groups. The disconnections on the pyridine ring **44** are the same as those in Corey's synthesis **29** but the chemistry used was quite different. Even the same strategy (same disconnections) need not restrict the chemist to one particular reaction.

So now we start the synthesis. Available **46** was converted in two steps into **45** and an interesting reaction with chloral hydrate and hydroxylamine gave the amido-oxime **47**, cyclised to **48** with polyphosphoric acid.[5]

This reaction makes the wrong ring size! But another interesting reaction with pyruvic acid makes the required quinoline **50**, esterified without isolation.[6]

The Reissert reaction would now require a base-catalysed acylation of the methyl group with dimethyl oxalate to give **42**. They say 'All attempts ... in the presence of various bases to produce intermediate [**42**] failed.' So they used the radical bromination of **43** with NBS (chapter 24) to give **51**, alkylated with the anion of methyl acetoacetate to give **52**.

The same kind of reaction as we saw in Corey's synthesis to give **39** now gave the hydrazone **53** and reduction finally produced the tricyclic compound **41**. As expected, oxidation was now easier, but still needed AgO and HNO₃. This is a linear sequence of 11 steps, but it was the first.

53; 70% yield   **41; 62% yield**

Convergence in a Commercial Synthesis

Convergence is even more important in commercial syntheses. The Merck anti-HIV drug MIV-150 **54** was synthesised in the laboratory in a 14-step linear sequence from *meta*-fluorophenol and **55** in 1% overall yield.[7]

54; Merck's MIV-150   ***m*-fluoro-phenol**   **55**

The fluorophenol could be converted into **56** in four good steps but the insertion of the vinyl group to give **57** by formylation and a Wittig reaction went in only 18% and the cyclopropanation with a diazoester and Cu(I) (chapter 30) gave poor selectivity in favour of the *cis* isomer of **57**. Worse still, it was necessary to protect the phenol as a methyl ether and the removal of the methyl group, the last step, went in only 52% yield, wasting nearly half of all the material.

56   **57**   **58; 45% *cis*-ester**

One branch of the convergent route made the ketone **59** while the other made enantiomerically pure *cis* cyclopropane **60**. These were combined to give the isocyanate **61** in two steps which was coupled to the amine **55** to give the urea **54**. Though protection was still needed and the last step was still poor, the overall yield was 27%—a considerable improvement on 1%. The details of the chemistry are too advanced for this book but appear in *Strategy and Control*.

59   **60**   **61**

Key Reaction Strategy

The Diels-Alder Reaction

We have already discussed strategies dominated by the availability of a starting material or the need to control stereochemistry. Another similar strategy revolves around one key reaction that achieves so much that it is worth basing our synthesis on it. The Diels-Alder is pre-eminent among such reactions. Some Diels-Alder disconnections are barely concealed by the structure of the target molecule: the sex pheromone **62** of the southern green stink bug *Nezara viridula* for instance. Getting back to the known Diels-Alder product **64** looks obvious.

It turns out that it is difficult to use the ketone to control the stereochemistry of the epoxidation. If acrylic acid is used as the dienophile, bromolactonisation of the product **67** gives a mixture of five- **68** and six-membered **69** lactones in 86% yield and a 1:1.5 ratio. Fortunately, treatment of both with an alkyl-lithium makes the same epoxide **70** by ring opening and S$_N$2 closure of the epoxide.[8] The reaction works well only with an electron-withdrawing group X such as SPh that must be removed later. Addition of MeLi to the ketone and elimination gives **62**.

The lycorine alkaloids come from plants such as daffodils. Among the simplest are the lycoranes **71–73** differing only in stereochemistry. They contain a saturated six-membered carbocyclic ring and that might make you think immediately of the Diels-Alder reaction.

71; α-lycorane   **72; β-lycorane**   **73; γ-lycorane**

Writing a general structure **74** for the lycoranes, we can remove the unique carbon atom between N and the benzene ring, with a Mannich reaction in mind, and disconnect the remaining C–N bond that does not go to a chiral centre to give **76**. For a Diels-Alder we need an electron-withdrawing group such as nitro **77** and an alkene in the ring such as **78**. Now Diels-Alder disconnection gives a simple diene **79** and a conjugated nitro-alkene **80**.

The regiochemistry **81** is fine: the nucleophilic end of the diene attacks the electrophilic end of **80**. But stereochemistry is all important. Undoubtedly the *trans* isomers *E*-**79** and *E*-**80** will be easiest to make so we should explore the result of the Diels-Alder with these two. We expect an *endo* transition state **82** and this gives **83**. Unfortunately this stereochemistry is wrong for all the lycoranes.

An alternative is to put the alkene in another place **84** and discover a different pair of diene **86** and dienophile **85**. Again the *E*-isomers will be easier to make and this time we get the right stereochemistry **88** for α-lycorane **71**. This strategy was followed in an early synthesis by Hill.[9] Another synthesis using the Diels-Alder reaction is by Irie.[10] More details appear in the workbook.

Aldol and Conjugate Addition Reactions

Among other important reactions that build up a molecule rapidly are aldol and conjugate addition. Together with Diels-Alder and Wittig reactions they are major players in organic synthesis. Another synthesis of lycoranes uses these reactions.[11] Similar preliminary disconnections with the addition of a carbonyl group **89** lead to the amino acid **90**.

Now the first conjugate addition, an intramolecular reaction between the nitroalkane and unsaturated ester in **92** is followed by conjugate addition of some aryl metal derivative to the unsaturated nitro-compound **93**. Both the unsaturated nitro-compound and ester could be made by aldol or Wittig reactions but there is clearly a potential selectivity problem.

The stable hemiacetal tetrahydropyranol **94** was used in a Wittig reaction to give the unsaturated ester **95** mostly as the *E*-isomer. Oxidation, nitroaldol and elimination gave the unsaturated nitro-compound **98**. It turns out that the aryl-lithium does conjugate addition without any copper and that it reacts exclusively with the nitroalkene to give **99**.

Now comes the key step: intramolecular conjugate addition of the nitroalkane anion to the unsaturated ester. When catalysed by CsF and a tetra-alkyl ammonium salt, this is selective (1.5:1) for the all equatorial products **100**. Reduction and cyclisation give the lactam **102** having the right stereochemistry for β-lycorane **72**.

Reductive removal of the amide carbonyl with borane and Mannich closure of the middle ring give β-lycorane **72**. A feature of this synthesis is that by changing the order of events and by adding ArLi with chelation control, all three lycoranes can be made selectively.

A Stereochemically Dominated Synthesis of γ-Lycorane

Stereochemistry has been a serious issue in all the syntheses of lycoranes but it dominates in one synthesis[12] of γ-lycorane **73**. All three hydrogens at the chiral centres are on the same face so the idea is that two of them could be put in by catalytic hydrogenation of a pyrrole such as **104** or **105** from the less hindered side of the alkene.

73; γ-lycorane   104   105

Another advantage of this approach is that we can now use electrophilic substitution on the pyrrole to add the rest of the molecule. So the secondary benzylic alcohol **106** might well cyclise to **105** with Lewis acid catalysis as the cation will be reasonably stable and the reaction is intramolecular. But the Friedel-Crafts alkylation to give **107** will not succeed as the cation would be primary.

105a   106   107

So the decision was taken to use succinic anhydride as the electrophile (chapter 5). Pyrroles prefer to react next to nitrogen with electrophiles (chapter 39), but with a large group on nitrogen **108** (i-Pr$_3$Si), Friedel-Crafts reaction occurred at the other position to give the keto-acid **109**. Reduction to the 'benzylic' alcohol and catalytic hydrogenation gave **110** in excellent yield.

108   AlCl$_3$ 109   SiR$_3$ 110; 95% from 108

After exchange of the protecting group, the Weinreb amide **111** (an alternative to nitriles for the formation of ketones discussed in detail in *Strategy and Control*) reacted with the aryl Grignard reagent to give, after reduction, **112**—the protected version of the alcohol **106** required for cyclisation.

111; 84% yield   2. NaBH$_4$ 112; 70% yield

Tin (II) triflate gave a quantitative yield of the Friedel-Crafts product **113**, emphasising the efficiency of cyclisation, and this compound was hydrogenated over platinum oxide to give only the required all-*cis* compound **114**. As there is an acyl group already on nitrogen, the original idea of using a Mannich reaction was replaced by intramolecular acylation with POCl₃ (Vilsmeier—chapter 39) and the amide could be reduced away with borane.

Notice that the three key reactions work brilliantly in this synthesis: the hydrogenation of **113** is totally stereoselective and very high yielding while the two electrophilic substitutions on the pyrrole are perfectly regioselective: acylation of **108** controlled by steric hindrance and alkylation of **112** controlled by electronic preference and because it is intramolecular.

We hope you will gather from these varied syntheses of one small group of natural products that even such relatively simple compounds can be made by a variety of strategies. There is no 'right' answer to a synthesis problem: workers in universities and industrial laboratories all over the world may devise routes to the same compound based on totally different reactions. In addition, solving the problems associated with a synthesis often brings into being new synthetic methods of lasting value.

References

1. S. T. Heller and S. R. Natarajan, *Org. Lett.*, 2006, **8**, 2675.
2. J. B. Hendrickson and J. G. deVries, *J. Org. Chem.*, 1982, **47**, 1148.
3. E. J. Corey and A. Tramontino, *J. Am. Chem. Soc.*, 1981, **103**, 5599.
4. J. A. Gainor and S. M. Weinreb, *J. Org. Chem.*, 1981, **46**, 4319.
5. C. S. Marvel and G. S. Hiers, *Org. Synth. Coll.*, 1932, **1**, 327.
6. A. R. Senear, H. Sargent, J. F. Mead and J. B. Koepfli, *J. Am. Chem. Soc.*, 1946, **68**, 2695.
7. S. Cai, M. Dimitroff, T. McKennon, M. Reider, L. Robarge, D. Ryckman, X. Shang and J. Therrien, *Org. Process Res. Dev.*, 2004, **8**, 353.
8. S. Kuwuhara, D. Itoh, W. S. Leal and O. Kodama, *Tetrahedron Lett.*, 1998, **39**, 1183.
9. R. K. Hill, J. A. Joule and L. J. Loeffler, *J. Am. Chem. Soc.*, 1962, **84**, 4951.
10. H. Tanaka, Y. Nagai, H. Irie, S. Uyeo and A. Kuno, *J. Chem. Soc., Perkin Trans. 1*, 1979, 874.
11. T. Yasuhara, K. Nishimura, M. Yamashita, N. Fukuyama, K. Yamada, O. Muraoka and K. Tomioka, *Org. Lett.*, 2003, **5**, 1123; T. Yasuhara, E. Osafune, K. Nishimura, M. Yamashita, K. Yamada, O. Muraoka and K. Tomioka, *Tetrahedron Lett.*, 2004, **45**, 3043.
12. S. R. Angle and J. P. Boyce, *Tetrahedron Lett.*, 1995, **36**, 6185.

Index

References to tables are given in bold type.

Printed and bound by CPI Group (UK) Ltd, Croydon, CR0 4YY